Stone Matrix Asphalt

Theory and Practice

Stone Matrix Asphalt

Theory and Practice

Krzysztof Błażejowski

CRC Press
Taylor & Francis Group
Boca Raton London New York

CRC Press is an imprint of the
Taylor & Francis Group, an **informa** business

Cover images: (Upper) SMA mixture with paver courtesy of Krzysztof Błażejowski; (Lower) finished asphalt pavement with SMA wearing course courtesy of Bohdan Dołżycki.

CRC Press
Taylor & Francis Group
6000 Broken Sound Parkway NW, Suite 300
Boca Raton, FL 33487-2742

© 2011 by Taylor and Francis Group, LLC
CRC Press is an imprint of Taylor & Francis Group, an Informa business

No claim to original U.S. Government works

Printed in the United States of America on acid-free paper
10 9 8 7 6 5 4 3 2 1

International Standard Book Number: 978-1-4398-1971-5 (Hardback)

Visit the Taylor & Francis Web site at
http://www.taylorandfrancis.com

and the CRC Press Web site at
http://www.crcpress.com

Contents

List of Figures

List of Tables

Foreword

Many years have passed since the beginning of stone matrix asphalt's (SMA's) worldwide success. Therefore now the right moment has come to review all accessible information.

Through this new book I have tried to assemble, in an organized manner, a certain body of knowledge obtained from a vast number of publications the world over, thus providing a review of the achievements of numerous engineers from various countries working on bringing recognition to SMA and developing its design methods. I did my best to explain that knowledge and to present it in an accessible way. Some useful hints resulting from my experiences encountered during the introduction of SMA in the early 1990s, discussions with many process engineers, and later reflections and observations round out the theoretical deliberations.

Knowledge of SMA is steadily improving, and new test results are revealed every now and then. In light of this, the moment when we can say that we really "know everything" about it is still far away. Alas, there is not one foolproof method for obtaining a perfect SMA in this book. Moreover, I tend to think such a recipe does not exist at all. This is not necessarily a bad thing because there is nothing like the ability to think and imagine when designing a mixture. The information put into the text is intended to help with using SMA. The range of accessible literature is really broad, so its accumulation and explanation are almost like putting together a jigsaw puzzle. Only the combination of many pieces of information enables associating certain relationships or finding out cause-and-effect relations. It is the reader's task to judge whether that jigsaw puzzle has been appropriately completed.

Undoubtedly, the diversified terminology and the methods of testing properties of constituent materials and mixtures were quite a challenge when compiling this book.

Summing up this foreword, I feel that it should be emphasized again that the knowledge of this technology constitutes the considerable sum of experiences of many people. Therefore there is no particular individual who knows "everything" about SMA. In other words, no matter how much you already know about SMA, it is always worth broadening your knowledge!

Krzysztof Błażejowski

Acknowledgments

If it were not for the support of the many people who played a part in the publication of this book, it could not have been completed. I would like to express my deep gratitude to all who helped me, in particular:

Prof. Klaus-Werner Damm, Germany
Dr. Bohdan Dołżycki, Poland
Lothar Drüschner, Germany
Horst Erdlen, Germany
Klaus Graf, Germany
Jan P. Heczko, UK
Konrad Jabłoński, Poland
Jiři Kašpar, the Czech Republic
Dr. Karol Kowalski, the United States and Poland
Janez Prosen, Slovenia
Gregor Rejewski, Germany
Halina Sarlińska, Poland
Marco Schünemann, Germany
Stanisław Styk, Poland
Ewa Wilk, Poland
Kim Willoughby, the United States
Bartosz Wojczakowski, Poland
Jan M. Voskuilen, the Netherlands

Also, I wish to express my sincere appreciation to Dr. Rebecca McDaniel, Purdue University, United States, for the time she devoted to the review and for a number of valuable contributions and suggestions that have substantially enriched the book.

The undoubtedly difficult-to-translate text has been quite a challenge for Leszek Mońko, Poland; and Murdo MacLeod, Scotland, who have been entrusted with its proofreading.

Author

Krzysztof Błażejowski, a graduate of the Civil Engineering Department of the Warsaw University of Technology, Poland, completed his PhD dissertation at the Kielce Technical University, Poland. Since 1992 he has been working as a research engineer for the Road and Bridge Research Institute in Warsaw, Poland, and then in research departments of various companies that manufacture such products as road binders, aggregates, and concrete. He is also the author of a series of publications on asphalt surfacing. In addition, Dr. Błażejowski remains active in standardization and training. When he is not involved in writing or research, Dr. Błażejowski spends time as a mountain guide.

Some Words on Terminology

Due to differences in terminology, chiefly between the United States and European countries, some assumptions were made. The universal term *binder* was used in the book instead of the U.S. term *asphalt cement* or the European *bitumen*. That decision has carried with it a change in the name from a binder course to an intermediate course.

Labeling mixtures according to the SMA 0/D system were used throughout the book where D denotes the nominal maximum particle size in a mixture. In Europe, marking SMA D (without '0/') is grounded in the standard EN 13108-5 and has been in use since 2006. Also, aggregate blends are labeled according to a similar system as d/D where d and D are the lower and upper limits of aggregate fraction, respectively; for example, a coarse aggregate 8/12.5 mm means aggregates with grains of size between sieves 8.0 and 12.5 mm with a permissible amount of oversizes and undersizes.

The abbreviations *m/m* and *v/v* refer to ratios by mass and volume, respectively. The abbreviation *PMB* means polymer modified binder.

1 The Concept of Stone Matrix Asphalt

For more than two decades, stone matrix asphalt (SMA), called stone mastic asphalt in Europe, has been taking over the global asphalt paving market at a remarkably high speed. Its fast-growing popularity has been surprising to many people. Although asphalt concrete may have appeared to be the indisputable leading choice for an asphalt layer, increasing vehicle axle loads have forced the application of new and better solutions.

1.1 A BRIEF HISTORY

The SMA mix, or *Splittmastixasphalt* in German, has been known since the mid-1960s. Dr. Zichner, a German engineer and manager of the Central Laboratory for Road Construction at the Strabag Bau AG, was its inventor. It was an attempt to solve the problem of the damage to wearing courses caused by studded tires. The trend in wearing course mixtures at that time in Germany was to use *Gussasphalt* (i.e., mastic asphalt) and an asphalt concrete with a low content of coarse aggregates. These types of surfaces were subject to fast wearing by vehicles equipped with studded tires. Both components, the mastic and the fine aggregates, were too weak to provide the mixture with suitable durability. Due to the high cost of pavement rehabilitation, strong demand for a new surface mix that could withstand studded tires was created. This was the impetus for Dr. Zichner's work.

When we try to picture ourselves in Dr. Zichner's situation, we would face—as he did—the task of designing an asphalt mixture that is resistant to wearing caused by studs and also is durable enough to have a long service life. He stated that coarse aggregate grains resistant to dynamic fragmentation, or crushing, were those that might guarantee the right wearing resistance. Thus they had to be the major components of prepared mixes to provide the needed wear resistance, whereas a high content of mastic and binder would produce a long service life. So the early idea for SMA consisted of creating a very strong aggregate skeleton of coarse aggregates and filling the spaces between them with mastic (i.e., a mix of binder, filler, and sand). That type of composition of an aggregate blend is typically called a gap-graded mineral mixture.

Initial trial attempts to construct the new mix consisted of spreading a hot mastic layer followed by spreading high-quality coarse aggregates over the mastic, then compacting the surface (with a road roller). The ratio of mastic to coarse aggregates (by weight) was 30:70. The mastic was made up of 25% B80 or B65 binder,* 35% filler, and 40% crushed sand (RETTENMAIER, 2009a). Based on experiments with such a composition, Dr. Zichner drew up a recipe for production of the mixture in an asphalt plant. The approximate composition of the first large scale production mix was as follows (RETTENMAIER, 2009a):

5/8 mm coarse aggregates	~70% (m/m)
0/2 mm crushed sand	~12% (m/m)
Filler	~10.5% (m/m)
B80 (B65) binder	~7.5% (m/m)

As we can see, there was none of the 2/5 mm aggregate in that mix that was typically used in other mix types; the absence of this size fraction produces a gap grading, as we shall see later (in Chapter 2).

Then the problem of draining-off the binder from the aggregate was encountered; with such a high binder content and few fines to hold the binder in the mix, the binder tended to flow off the coarse aggregate particles. It was acknowledged, after laboratory tests, that an additive of asbestos fibers would be a good drainage inhibitor (so-called stabilizer). Such a designed mix could be produced, transported, and laid in a traditional way (RETTENMAIER, 2009a).

The mixes were named by Dr. Zichner in 1968 as follows (Zichner, 1972):

- MASTIMAC—the name referring to mixes for layers 2–3 cm thick,
- MASTIPHALT— the name referring to mixes for layers thicker than 3 cm.

Early road sections of MASTIMAC were used on internal roads of asphalt plants belonging to the Strabag/Deutag Consortium, enabling them to gain experience with the new mixture. Eventually a public road was paved with the MASTIMAC mix on July 30, 1968, in Wilhelmshaven, Germany, on Freiligrath Straße. The result was so encouraging that some other sections were paved with MASTIMAC soon afterward (RETTENMAIER, 2009b). Gradation curves of the new mixtures (Figure 1.1) were presented in a German publication (Zichner, 1972).

The stone mastic composition and its laydown were patented by Dr. Zichner in Germany, the United States, Sweden, France, and Luxembourg.[†] It is interesting how the inventor described his ideas in the U.S. patent text (Zichner, 1971):

the gravel size mixtures are composed…so that the percentage of the coarser size is greater than that of the smaller size. In this manner a relatively great interstitial volume is achieved in the gravel mix on the one hand, and on the other hand a good interfitting

---

* Road binders (bitumens) B65 and B80 were based on Pen@25°C range according to German DIN standard: B65 was 50–70 dmm, B80 was 70–100 dmm.
[†] Germany, Patent No. 1926808 (1969); United States, Patent No. 3797951 (1971); Sweden, Patent No. 7110151-3 (1972); France, Patent No. 71.28874 (1971); Luxembourg, Patent No. 63688 (1971).

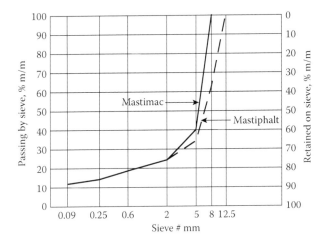

FIGURE 1.1   Grading curves of Dr. Zichner's mixtures: Mastimac and Mastiphalt. (Based on Zichner, G., MASTIMAC unad MASTIPHALT bituminöse Gemische für hochwertige Deckschichten. STRABAG Schriftenreihe 8, Folge 4, 1972.)

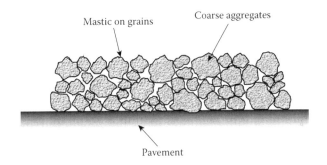

FIGURE 1.2   First stage of the mix performance according to U.S. patent No. 3797951—after laying.

of the individual pieces [of] gravel is assured....The quantity and fluidity of the mastic is such that during and after the compacting, the mastic flows into the interstices between the stones forming the wearing course as aforesaid, and which the volume of the mastic is less than the interstitial volume of the stones....

In that patent, the approximate percentage of the mix composition was defined as 70% coarse aggregates, 12% filler, 8% binder, and 10% crushed sand. It was also indicated that stabilizing additives may be needed as well. As the reader can see, the mixture described above is similar to the contemporary understanding of an SMA mixture (Figures 1.2 and 1.3).

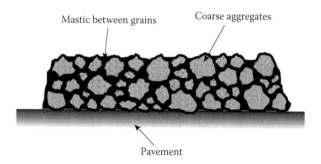

FIGURE 1.3   Second stage of the mix performance according to U.S. patent No. 3797951— after finishing.

Today it is generally admitted that the idea of SMA has changed little since its inception. What is SMA today? We can agree on its definition here; SMA is an asphalt mixture containing a gap-graded aggregate mixture, with high contents of coarse aggregate fractions, filler, and binder. Most often a stabilizing additive (drainage inhibitor), which prevents the draindown of the binder from the aggregate, is needed.

Although SMA mix was originally intended only for wearing courses, in some countries it is also applied in binder (intermediate) courses (see Chapter 13). Polymer modified binders were not commonly available in the 1960s and 1970s, so only conventional binders were used. Although the binders were quite soft, it was soon noticed that SMA layers were very durable, and their rut resistance became especially evident.

The SMA mix was not forgotten in 1975 when a ban on the use of studded tires was introduced in Germany (Bellin, 1997). The concept of SMA turned out to be good not only for that kind of damage but also for rutting resistance and durability. Today SMA is regarded as an ideal mixture for heavy-duty asphalt pavements that require the highest resistance to damage and a very long service life. It is a practically synonymous with resistance to rutting.

1.2   FROM GERMANY TO…OR SMA ALL OVER THE WORLD

Until the beginning of the 1980s, SMA was essentially known only in Germany. Its application in other European countries was limited in scope. Scandinavian states where studded tires used were the quickest to adopt the SMA concept; for example, in Sweden a few sections of roads had been paved with SMA by 1974 European Asphalt Pavement Association (EAPA, 1998). In Poland, which at that time was behind the Iron Curtain, concepts in West German publications were officially disallowed. Despite that, the Polish road administration permitted the first road section of an SMA-like mixture to be placed within its borders in 1969 (Jabłoński, 2000). Very positive results of that trial made the Polish Central Authority of Public Roads inclined to draft a standard (ZN-71/MK-CZDP-3), which was put into practice in 1971. After forschungsgesellschaft für straβen-und verkehrswesen (Germany)

(FGSV) published the first German technical standard for SMA (ZTV bit-StB 84), the mix became more popular and several European countries started to test the SMA mix. Now, virtually all European countries use SMA or, like France, nationally standardized mixtures conceptually similar to SMA.

The significant growth of the SMA application started in the early 1990s outside of Europe. This growth was certainly boosted by its popularity in the United States, and research on developing an American method of designing an SMA mix commenced (see Chapter 7). Popularization of SMA in North America led to the release of SMA guidelines in other countries such as Australia, New Zealand, and China. During the last 20 years, SMA has become a global mix, and thus it may be seen almost everywhere where mineral-asphalt layers are applied.

1.3 STRENGTHS AND WEAKNESSES OF SMA

SMA owes its fast and wide-spreading growth to some unquestionable merits such as the following:

- Long working lifetime (service life)
- High resistance to deformation due to the high-coarse aggregate content and strong skeleton of interlocked aggregate particles
- Increased fatigue life as a result of the higher content of binder
- Increased in-service traffic wear resistance because of the presence of hard coarse aggregate grains,
- Good macrotexture of the layer surface and decreased water spray generated by traffic on wet surfaces
- Good noise-reduction properties

However, despite its strong points, the following drawbacks also exist:

- Low initial skid resistance unless a fine aggregate gritting or a crushed sand finish is applied
- High cost of the mix compared with a conventional asphalt concrete (initial costs can be increased by 10–20% due to higher contents of binder, filler, and stabilizer, but the extended service life of the pavement can result in reduced life cycle costs)
- Risk of different types of fat spots appearing on the surface as a result of errors or variability during SMA design, production, or construction

To end this part of the discussion, let us recall the words of Dr. Zichner of his patent:

All in all, the method taught by the invention and the procedures made possible thereby provide a wear surfacing characterized by high resistance to abrasion, long-lasting roughness, and reliable adhesion to the road surface. (Zichner, 1971).

1.4   DIFFERENCES IN SMAs AROUND THE WORLD

To summarize the status of SMAs around the world in regard to the design of mixtures and their aggregate gradations, two general trends consist of the following:

- German SMAs and others made by those more-or-less faithful to German guidelines have evolved somewhat based on long-standing systematic observations and experiences in SMA technology
- Research and development continues on new ways of designing SMAs; U.S. and Dutch techniques may be representative of that trend (see Chapter 7).

Various mixtures worldwide are referred to as SMA although they may differ greatly from Dr. Zichner's invention. Some of these variants should, in principle, actually be called something besides SMA. Undoubtedly they are still gap-graded mineral mixes and SMA-like, but they are not identical with Dr. Zichner's *Splittmastixasphalt*.

In subsequent chapters, SMA composition will be discussed according to a scheme drawing on both trends. The existence of so many different attitudes toward the SMA mix has been problematic in selecting the right way to express the variety of opinions in a methodical and comprehensive way. That is why some specific references to particular trends may appear occasionally.

1.5   CONTENT OF SMA

The content of SMA will be divided into the following parts (Figure 1.4):

- Coarse aggregate skeleton
- Mastic (i.e., binder, filler, fine aggregate, and stabilizer)
- Voids in the asphalt mix

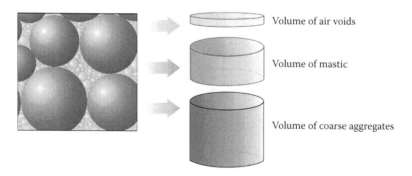

FIGURE 1.4   Division of SMA into basic components.

This division of SMA into parts (with the predominance of the main two—skeleton and mastic) has been adopted to better explain the roles of each component of the SMA mix. A similar division may be found as a directly applied approach in some designing methods (e.g., the U.S. and Dutch methods).

The SMA components of the aggrate skeleton and mastic will be discussed in Chapters 2 and 3, respectively, and the significance of voids in the SMA mix will be discussed in Chapter 6, when we look at designing SMA, and in Chapter 9, when we examine the laydown (placement) of SMAs.

Now it is time to tackle the subject on hand!

2 Skeleton of Coarse Aggregates

In this part of the book we shall deal with the grains within the mix structure that are active in forming a coarse aggregate skeleton. The following significant questions will be answered, too:

- How is a stone matrix asphalt (SMA) mix skeleton formed?
- What does a gap gradation mean?

2.1 DEFINITION OF AN SMA AGGREGATE SKELETON

The notion of an *aggregate skeleton* of a mixture has a pretty broad meaning. Figure 2.1 shows various types of mineral mixes with different interactions of grains. These range from a sand mix, through mixes where the coarse aggregate particles occupy a more and more substantial share of the volume, and up to a mix consisting entirely of coarse aggregates. The types of aggregates that make up a continuous matrix and form a load-carrying component determine the specific group ranking.

While mixes consisting mainly of an aggregate with sizes up to 2 mm (or 2.36 mm in the United States) (e.g., sand asphalt) are rarely applied, sand–coarse aggregate mixes are within a continuum from mastic asphalt to asphalt concrete. SMA belongs in a coarse aggregates–sand group with porous asphalt, which is the type closest to the straight coarse aggregate type. Mineral mixes with exclusively coarse aggregates make so-called coated macadams, which are seldom used at present.

Now we deal with the first component of an SMA—the skeleton formed of coarse aggregates. Then in Chapter 3 we shall discuss the mastic made of a filler, sand, and binder.

First let us define the term *coarse aggregate skeleton*. A coarse aggregate skeleton is a structure of grains of suitable size that rest against each other and are mutually interlocked. In Europe, the coarse aggregates are generally taken to be those larger than 2 mm. Following the U.S. option, let us adopt an assumption that a fixed size of 2 mm (2.36 mm in the United States) is not necessarily the lower limit of the coarse aggregate fraction. Why? Well, to meet the structural requirements of a particular grain arrangement (layout), the mechanical resistance of the aggregate blend must be high enough to withstand loads. And this is related to the size of grains, among other factors. So let us agree that a skeleton is made up of adequate size grains but their lower limit is not necessarily equal to 2 mm (2.36 mm). Since the skeleton consists of various coarse aggregate grains, we often use the more universal term *coarse aggregate skeleton*.

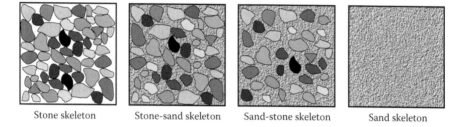

| Stone skeleton | Stone-sand skeleton | Sand-stone skeleton | Sand skeleton |

FIGURE 2.1 Division of mineral mixes into types depending on interactions between sand grains and coarse aggregates. (From van de Ven, M.F.C., Voskuilen, J.L.M., and Tolman, F., The spatial approach of hot mix asphalt. Proceedings of the 6th RILEM Symposium PTEBM'03. Zurich, 2003. With permission.)

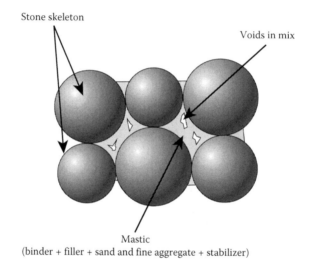

Mastic
(binder + filler + sand and fine aggregate + stabilizer)

FIGURE 2.2 Division of SMA into basic components: coarse aggregate (skeleton) and mastic.

To classify all grains of the mineral mix, they may be divided as follows (Figure 2.2):

- Those forming a skeleton (*skeleton makers*) and carrying loads
- Those filling in the voids in a skeleton and not carrying loads

That division also coincides with the following frequently used terms:

- Active grains (i.e., those forming a skeleton)
- Passive grains (i.e., those filling in voids)

The selection of a sieve to determine a *skeleton maker* is of great importance for the properties of a newly designed mix. That matter will be treated in detail in Chapter 6.

2.2 FORMATION OF A COARSE AGGREGATE SKELETON

What is the reason for developing mixes with stronger mineral skeletons? Surely it is because of heavier and heavier traffic. Not only have axle loads and traffic volumes grown steadily, but the structures of vehicle tires have also changed. The increased popularity of super single tires, for example, has changed the level of stresses applied to pavements. Obviously all those factors magnify the requirements for asphalt mixtures.

A well-known example can be used to present the SMA mineral skeleton structure. If we put some coarse grains in a pot (Figure 2.3), compact and then load them, we shall obtain a structure with high compressive strength, depending on the aggregate's fragmentation (crushing) resistance. A distinctive feature of such a compacted collection of coarse grains is the full and uninterrupted contact between them. This type of skeleton might be desirable for an asphalt surface mixture to provide a strong structure.

Now let us look at Figure 2.4, which presents a schematic showing how a load is carried by a mineral skeleton (of a surface course), assuming full grain contact. The transfer of load by adjacent grains through contact points between the coarse particles may be seen. If these contact points between coarse aggregate particles are not present, the finer particles will have to help carry the load, which results in the

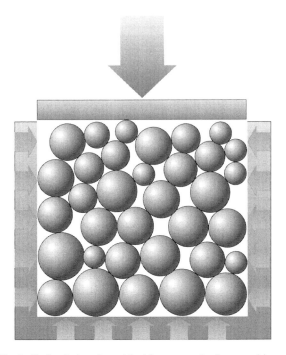

FIGURE 2.3 Vertically loaded grains with side support (as in a crushing resistance test of coarse aggregates).

(a)

(b)

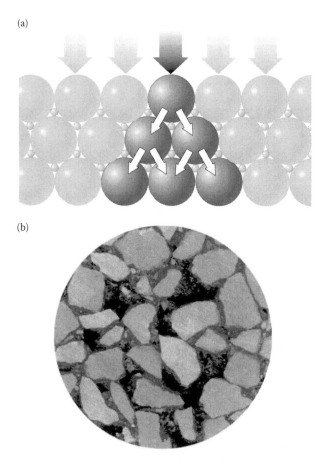

FIGURE 2.4 Load distribution among balls in case of a uniform load distribution: (a) schematic; (b) cross section of a Kjellbase sample: asphalt mixture consisting of a one-fraction coarse aggregate with a minimal quantity of passive aggregate, which is an exemplary application of full gap gradation for making up a strong skeleton (see Section 13.4.1) (From Sluer, B.W., Kjellbase. A future without ruts. Development of a heavy duty pavement. Presentation at the 1st International Workshop. SMA and JRS Fibers. Hannover, 2002. With permission.)

development of a discontinuous load transfer by coarse, active grains and potentially weakens the whole structure.

Looking at Figures 2.3 and 2.4, we may notice one of the SMA's characteristic features. During the compaction process on a construction site, the aggregate grains in the SMA skeleton are forced into direct contact. The coarse grains are brought into contact with each other, making the desired skeleton. Once that contact occurs, further compaction may be harmful. Why? Because it will crush grains. Let us look once again at Figure 2.3. Since the skeleton has already come into existence, further compacting will only lead to crushing grains. In other words, an SMA mixture has to be skillfully compacted in such a way that the coarse grains are properly placed, securing stone-to-stone contact. This principle applies to compacting energy on a construction site, as well as in a laboratory.

To conclude our examination of the formation of an SMA skeleton, it is worth noting an idea put forth by Van de Ven et al. (2003), who said that there is probably no real, 100% stone-to-stone contact in a newly compacted SMA and that the coarse grains are separated from each other by the finest grains of filler, sand, and thin binder film. This means that there is an *enlarging effect** of the volume of voids in the stone skeleton. The final arrangement of grains comes after some time under the influence of traffic and temperature of the layer. Small grains (sand and filler) can be crushed or moved, and the void content in SMA decreases.

Let us also bear in mind that the more stable a mineral skeleton is, the less susceptible it will be to deformation. Even when the binder softens due to the increased temperature of the pavement, the layer will not necessarily be deformed if there is an adequate blend of aggregates. The weaker the skeleton and the higher the temperature, the greater is the role of the mastic's shear strength, and the more reason for reinforcing the mastic with polymers or special fibers.

2.3 GAP GRADATION

Our aim in designing an SMA's aggregate structure has already been identified—a strong skeleton of coarse grains. Let us now consider what requirements an aggregate mix has to meet to create such a desirable skeleton. There is no room in it for too many fine or weak grains. The key solution for that question is gap gradation—that is, the right proportion of grains of defined sizes.

Let us start by examining a continuous gradation. If we want to design an aggregate mix with a maximum density (or otherwise, with a minimum void content), we should create it from an aggregate with a roughly equal share of grains from consecutive fractions. In other words, such a mix should contain a proportionally even quantity of all fractions. We would call this type of gradation a continuous gradation. The appearance of grains of different sizes makes closer packing in a volume unit possible. This also minimizes the volume of voids among grains. Asphalt concrete is an obvious example of a continuously graded mixture (Figure 2.5; solid line).

So what is a gap gradation? Gap gradation is a disruption in the occurrence of consecutive aggregate fractions in an aggregate blend; that disruption results from a lack or minimal amount of one or more aggregate fractions. Looking at Figure 2.4, we can see the formation process of a skeleton with coarse grains and some of the finest grains but without the sizes in between. **Gap gradation means a lack or minimal share of specified fractions of intermediate aggregates**. The role of the gap gradation is so essential that the lack of definite size grains must be evident. But which fraction or sizes of grains or fractions? Here we have a couple of definitions and methods.

---

* This subject will be discussed in Chapter 7.

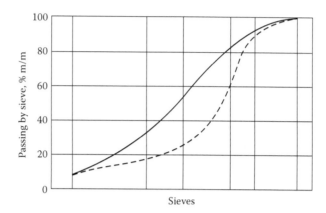

FIGURE 2.5 Position of gradation curves of aggregate mixtures: solid line, asphalt concrete (continuously graded); dotted line, SMA (gap graded).

2.4 DEFINITION OF AN SMA SKELETON ACCORDING TO THE ORIGINAL GERMAN METHOD

The original German approach (by Dr. Zichner) to designing SMA aggregate blends is based on having adequate ratios of various aggregate fractions. In that context, the stone-to-stone contact is neither specifically analyzed nor controlled.

Since the beginning of the SMA concept, the ratios established by Dr. Zichner have been only slightly changed. Thus we may emphasize clearly that Dr. Zichner's mixture is a really genuine SMA. The contemporary weight ratios of coarse aggregate fractions preferred in Germany (Drüschner and Schäfer, 2000) are shown in Table 2.1.

In the German design of the SMA coarse aggregate skeleton, all fractions of aggregate bigger than 2 mm are used. For example, for an SMA 0/11 we take not only aggregate sizes 8/11 and 5/8 but 2/5 mm as well. Manipulation of the ratios of these different sizes is required (see Table 2.1) to provide the desired skeleton by minimizing the share of 2/5 mm aggregate down to one part in seven (15%) of coarse aggregates' mass fraction. Consequently the German SMA gradation curves have no sharp breaks related to the absence of successive aggregate fractions and do not exclude any fractions larger than 2 mm (i.e., all fractions of coarse aggregates are used). In a way, most of the original German gradations are *quasi*-continuous gradings (all fractions are present in the mix), with a minimal share of specific coarse aggregate fractions. It should be emphasized that in most German guidelines the amount of coarse aggregates (bigger than 2 mm) are not very high. For example, for SMA 0/11 the lower limit of this fraction has been changed from 75% (in 1984) to 73% (in 2001). There were many reasons for this change, including better compactability, lower permeability, and improved rutting resistance.

The success of SMAs designed according to the ratios of aggregate sizes presented in Table 2.1 has been proven through long-term pavement evaluations, although these ratios have been refined now and then (within a limited scope). So the strength of the

TABLE 2.1
Weight Ratios of Coarse Aggregate Fractions in SMA[a]

SMA Type	2/5 mm	5/8 mm	8/11 mm
SMA 0/8	2.5 parts	4.5 parts	NA
SMA 0/8S	2 parts	5.5 parts	NA
SMA 0/11S	1 part	2 parts	4 parts

Source: Drüschner, L. and Schäfer, V., Splittmastixasphalt. DAV Leitfaden. Deutscher Asphaltverband, 2000. With permission.

Note: NA = not applicable; S = mix for heavy traffic.

[a] Based on the German DAV Publication.

German method is a designed SMA skeleton composition based on long observation and experimentation. (See Chapter 7 for a detailed description of the method used in Germany.)

2.5 DEFINITION OF AN SMA SKELETON WITH OTHER METHODS

In some countries (e.g., the United States and the Netherlands) a method of constructing an SMA skeleton has been developed based on the control of stone-to-stone contact or a real gradation discontinuity. Based on these methods, the definition of SMA is expanded to mean an asphalt mixture containing mastic stabilizer (drainage inhibitor) with a gap-graded aggregate blend and a very high content of coarse aggregates in which smaller grains are seated among the bigger ones, filling voids among them but not shoving them aside. Based on this definition of SMA, it is necessary to determine the level of gradation discontinuity at which active grains are not shoved aside by passive ones.

Let us look at Figure 2.6a to d, which shows an idealized arrangement of grains (represented by smooth spheres) in an aggregate blend and illustrates the relationships between the radii of the active coarse grains (marked R) that form the skeleton and the radii of the passive fine grains (marked r). The proportions of the different grain sizes have been selected so that the smaller grains do not shove the bigger ones aside.

Thus we have two sets of spheres, here being examined two-dimensionally. Now we may theorize a bit on their significance for the sought-after discontinuity.

- When arranging spheres and circles as in Figure 2.6a and b, the passive grains cannot be bigger than $0.41 R$. So, in an SMA 0/12-mm mixture. Assuming that the active grains are 8/12 mm, the next smaller size fraction should be just 2/5 mm to prevent shoving the active grains apart. The desired gradation discontinuity is developed by the absence of the 5/8 mm fraction.

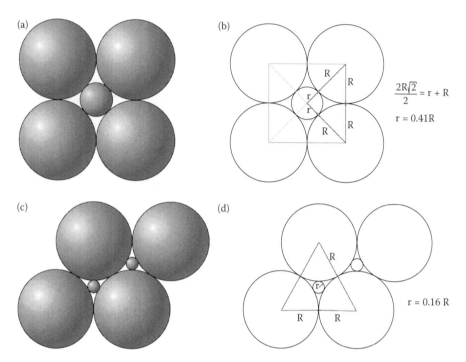

FIGURE 2.6   Various sets of spheres (and circles) and sizes of passive grains in relation to active ones.

- Such an arrangement of grains as shown in Figure 2.6a and b is unnatural and unlikely. The one presented in Figure 2.6c and d is more likely. But it is easy to see that one consequence of a better (closer) distribution of coarse grains is the reduction of free space available for filling (passive) aggregate. Simple geometric analysis enables the calculation of the maximum dimension of fine grains, equal to 0.16 R, to avoid shoving coarse grains aside in this scenario. If we look again at the example of the SMA 0/12 mm, apart from the active 8/12 mm fraction, the next material will be just the passive 0/2 mm (sand) fraction.* In the arrangement illustrated in Figure 2.6c and d, the necessary gradation discontinuity would consist of the absence of the 2/8 mm fraction.

These deliberations have been carried out on the assumption that the only active fraction is the 8/12 mm one, but this is by no means obvious. The SMA skeleton may also involve active grains smaller than the 8/12 mm fraction, such as grains bigger than 5 mm. Active grains of the 8/12 mm and 5/8 mm fractions would have a considerable influence on the formed skeleton owing to the following:

---

* For more information, see the discussion of Kjellbase in Chapter 13.

- Better packing of coarse grains (here, larger than 5 mm)
- Reduction of the size of voids and therefore the size of the filling grains

If the 5/8 mm fraction is also active, the desired grain discontinuity may be secured by the absence of the 2/5 mm fraction. The Bailey method, described in detail in Chapter 7, is based on a similar geometric calculation; that approach assumes that each successive aggregate fraction should be equal to about 1/5 (0.22) of the larger fraction. This requirement is intended to prevent smaller particles from shoving the coarse aggregate skeleton aside. As we can see, the path from geometry to design method is not a long one.

These considerations are only theoretical because in a real mixture we do not deal with balls and the layout of grains is different. Yet those examples point out a significant issue; interfering with the gradation discontinuity (by adding aggregates of that size fraction) causes the loss of stone-to-stone contact of the coarse grains that form the SMA skeleton (Figure 2.7). This conclusion forms the basis of the U.S. method of SMA design and the so-called Bailey method (both presented in Chapter 7). The same conclusion has also been used in some Scandinavian solutions (e.g., in the *real SMA* concept).

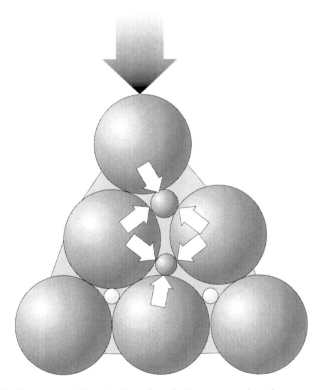

FIGURE 2.7 Interrupted discontinuity of gradation occurs when the presence of the in-between size grains disturb the action of the coarse aggregate skeleton.

A final remark—looking at Figure 2.4, we may be inclined to make a mixture of only one size fraction of coarse aggregate (8/12 mm) with mastic (0/2 mm). Obviously we would get a very strong aggregate skeleton, but we cannot forget the following technological consequences:

- Placement would be difficult; the coarse grains might be pulled by the screed plate.
- Compaction would be complicated; such a mixture is difficult to compact.
- We may have difficulty keeping the void content below 6% (v/v)* in the compacted surface course.

These reasons demonstrate the need for smaller, but still coarse, grains to participate in forming the skeleton. Further discussion on the features of such a mixture and analysis of the influence of the coarse aggregate gradation will be found in Chapter 6.

2.6 COMMENTS ON THE PRESENTED METHODS

A downward trend that has been recently evident in Germany involves reducing the aggregate blend gradation discontinuity and establishing an upper limit on the coarse aggregate (bigger than 2 mm) content. These changes have been justified as ways to secure a more durable wearing course and to improve its compaction.

Although in some respects very appealing, another viewpoint on building a very strong SMA skeleton with coarse grains only is not without controversy. From a technological point of view and also from Dr. Zichner's original principles, SMA is almost a gap-graded mixture. But are all gap-graded mixtures SMAs? Certainly not. If something is going to be called SMA, namely *Splittmastixasphalt*, its distinctive feature should be a gradation with ratios at least similar to those listed in Table 2.1.

The evolution of requirements for SMA gradation is also apparent in the United States. Some U.S. guidelines require SMA mixtures marked by very distinct gap gradation while other guidelines do not differ much from the German regulations. This trend may be well illustrated by comparing the SMA 0/12.5 mm gradation curves according to the American Association of State Highway and Transportation Officials (AASHTO)[†] M 325-08 standard with those from the National Asphalt Pavement Association Quality Improvement Series No. 122 (NAPA QIS-122) guidelines, as shown in Figure 2.8. It may be easily noticed that SMA gradations based on NAPA QIS-122 may be marked by substantially more single-sized aggregate and a stronger skeleton of 9.5- to 12.5-mm grains (up to 73% of grains retained on a 9.5-mm sieve), while gradation curves according to AASHTO M 325-08 look more continuous.

---

* It is a requirement regarding an SMA surface course after its incorporation; for more information, see Chapter 10.

[†] AASHTO is the standards setting organization; founded in 1914, it was known as the American Association of State Highway Officials (AASHO) before 1973.

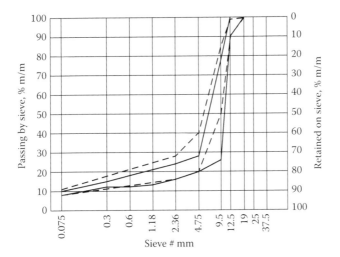

FIGURE 2.8   The SMA 0/12.5 mm gradation curves of the United States. Solid lines, gradation after NAPA QIS-122; dotted lines, gradation after AASHTO MP 8-00. (Based on data from AASHTO MP 8-00 and NAPA QIS 122. Designing and Constructing SMA Mixtures— State of the Practice. National Asphalt Pavements Association, Quality Improvement Series 122. 2002.)

2.7   AMONG THE SKELETON GRAINS

Having formed an adequate skeleton of coarse grains, we have to remember to place mastic between the active grains. Achieving the proper volume of mastic is critical; there must be the right amount of mastic to coat the coarse grains but at the same time to leave some free, unoccupied space. Figure 2.9 illustrates one way to look at the packing of SMA by reflecting the volume contents of coarse aggregates, mastic, and voids.

It may be concluded from Figure 2.9 that the volume of voids among the coarse aggregates has to be properly determined at the aggregate skeleton design stage. The ideal laboratory design method is one that accurately defines the volume of the free space between the coarse aggregates as they would exist after field compaction.

The design method may not accurately reflect the mixture volumetrics after field compaction, especially if the field compaction is a more effective compaction than assumed at the SMA design stage. This would result in a mastic volume that is too big in relation to the free space among the coarse aggregates, and as a result, mastic may be squeezed up to the surface,* causing fat spots to appear. It is also interesting that in such circumstances an unexpected decrease of free space for the mastic may result. But that is the subject of the following chapters.

Thus we have reached the moment when we have to deal with the mastic.

---

* See Chapter 11.

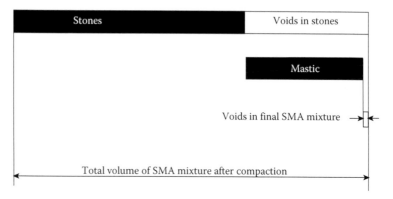

FIGURE 2.9 Volumes of SMA coarse aggregate skeleton and mastic. (Modified from Voskuilen, J.L.M., Ideas for a volumetric mix design method for Stone Mastic Asphalt. Proceedings of the 6th International Conference Durable and Safe Road Pavements, Kielce (Poland), 2000. With permission.)

2.8 SUMMARY

- SMA mixtures belong to a group of coarse aggregate–sand mixtures with a continuous coarse grain matrix—that is, their skeletons are formed by interlocked coarse aggregate particles that transmit loads.
- The term *coarse aggregate skeleton* is a conventional notion meaning a structure made up of grains with a specified lower limit of size. The most frequently accepted limit is the 2-mm (or 2.36-mm) sieve, though in many countries, such as the United States, it depends on maximum grain size.
- The SMA aggregate grains may be divided in two types: those forming a skeleton and carrying loads (so-called *active* grains) and those filling voids and carrying virtually no loads (so-called *passive* grains). The basis for developing an active grain skeleton in an SMA is to ensure contact exists among the active grains.
- A discontinuity in the SMA aggregate blend gradation (gap gradation) is produced by the lack of in-between fractions of aggregate bigger than 2 mm.
- A coarse aggregate skeleton as defined by the German method (the original method, after Dr. Zichner's idea) consists of an adequate proportion of all aggregate fractions greater than 2 mm.
- A coarse aggregate skeleton defined by some other methods, such as the U.S. method, consists of an aggregate blend with a decisive predominance of coarse aggregate and a distinct gap gradation, which results in establishing stone-to-stone contact.

3 Mastic

Mastic is the second largest component of stone matrix asphalt (SMA); it is approximately 20–25% by weight of the mixture and 30–35% by volume. About 35–40% (v/v) of the compacted coarse aggregates is made up of voids, and after filling the aggregate with mastic, 3% to 5% (v/v) of empty space will be left.

Mastic* consists of the following:

- Fine aggregate
- Filler
- Stabilizer (drainage inhibitor) in the form of fibers or other additives
- Bituminous binder

Filler and binder make up mortar. Blends of the fine fraction of the filler with binder act like binders and can be tested as a binder, but blends of the total filler and binder act more like a mixture and can be tested in that manner (e.g., BBR† stiffness, resilient modulus, and tensile strength) (Brown and Cooley, 1999).

In the preceding chapter we discussed how coarse aggregate particles make up a skeleton. The task of the coarse aggregate skeleton is different from that of the mastic. The functions of mastic are as follows:

- Binding (sticking together) the coarse aggregate skeleton
- Lubricating the coarse grains during compaction and enabling a proper aggregate structure in a compacted surface course
- Sealing the layer, or filling the voids in the compacted aggregate structure to provide it with high durability and resistance to other external factors such as water or deicers
- Withstanding stresses caused by load and temperature

Figure 3.1 shows the close packing of fine (passive) aggregate among coarse (active) grains.

Now let us deal with the mastic components.

3.1 FINE AGGREGATE

Throughout this book the term *fine aggregate* has been conventionally used as a term for the passive aggregate. Its upper limit depends on us—or more specifically, on

---

* In this book definitions of the mastic and mortar will be used as presented here. However, in numerous countries these definitions differ from those adopted in this book; for example, in the United States mastic is called total mortar and mortar is called fine mortar (and contains drainage inhibitor, which is a stabilizing additive).
† BBR - Bending Beam Rheometer.

(a)   (b)

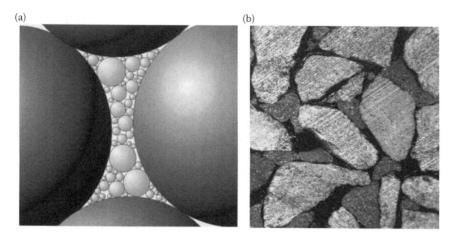

FIGURE 3.1 Filling voids among coarse grains of aggregate with passive grains: (a) schematic diagram and (b) an actual example—mastic marked with dark gray color. (Photo courtesy of Krzysztof Błażejowski.)

the type of coarse grain previously in process of SMA design accepted as a skeleton maker. The task of the fine aggregate is to fill voids among the coarse aggregate particles and facilitate their interlocking, though it is likely to be put the other way round; the fine aggregate cannot disturb interlocking of the coarse aggregates. Such a disturbance can best be illustrated by an example of rounded, uncrushed aggregate (e.g., natural sand or uncrushed gravel) with smooth surfaces that allow the coarse aggregates to glide easily. Introducing such "hard balls" into an SMA mixture causes problems with the stabilization (interlocking) of the aggregate skeleton. That is why in many countries the incorporation of uncrushed aggregates in SMA has been limited to only low volume roads, or its use has been generally banned.

Angularity is a feature that describes the properties of fine aggregates and is typically defined in terms of a flow test, which is an indirect method of angularity measurement. In Europe, the flow coefficient (method EN 933-6) is labeled with the *Ecs* symbol according to European Standard EN 13043 and describes the time necessary for a standard amount of aggregate to flow out of a suitably shaped vessel through an opening. Obviously the more crushed aggregate with a better microtexture, the longer the flow-out time; that is, rough or angular aggregates tend to lock up and not flow as quickly as smoother particles. Aggregate with an *Ecs* greater than or equal to 30 or 35 seconds (i.e., *Ecs*30 or *Ecs*35 category) is regarded as appropriate for SMA. There are also requirements concerning the angularity of fine aggregate applied in the United States.[*] Tests are carried out in accordance with the American Association of State Highway and Transportation Officials (AASHTO) T 304 Method A (ASTM C1252), and the required fine aggregate angularity (FAA) values are greater than or equal to 45% (NAPA QIS 122). Angularity can also be

---

[*] The flow rate method, similar to the European concept, is not standardized in the United States Comparison tests have been conducted and are published in (Tayebali et al, 1996).

measured according to ASTM D3398 and a National Aggregate Association (NAA) test method as well.

Remembering the positive influence of angularity on properties such as deformation resistance, we still have to take into account some negative factors like compaction resistance that accompany an increase in aggregate angularity. Moreover, research conducted by Stakston and Bahia (2003) showed that the effect of FAA depends on the source aggregate and its gradation and that angularity could have an adverse effect on a mixture's resistance to shearing. Further reading on the influence of FAA on asphalt mixtures can be found in the Stakson and Bahia study or in other papers (Johnson et al., 2004; White et al., 2006).

The content of free voids in a compacted fine aggregate is undoubtedly an essential parameter in SMA volumetric design. This characteristic may be tested through various methods, (e.g., the AASHTO T19 standard). The packing of consecutive SMA ingredients cannot be determined without prior knowledge of this characteristic. That approach to SMA design is discussed in Chapter 7 in the section devoted to U.S. and Dutch design methods.

3.2 FILLER

The term *filler* means an aggregate that mostly passes through a specified sieve (0.063 mm in Europe, 0.075 mm in the United States). It should be emphasized that the material just discussed, which is generally called filler, denotes all the grains—that is, both those coming from the added filler and those occurring on fine and coarse aggregate grains in the form of dust. Thus if we want to know the behavior of a filler fraction in a given mixture, then all the grains below a specified sieve size in a final mineral blend should be separated, regardless of their source. All that material should be tested. If we only test the added filler, the results do not show the influence of the entire filler fraction on the properties of the mix.

The significant influence of filler on asphalt mixtures may be defined in the following way (Anderson et al., 1982; Kandhal et al., 1998; Drüschner, 2006):

- Filler grains smaller than the bitumen film on aggregates can behave like a carrier (binder extender); very fine filler makes the mix behave as if there is even more binder present, which may result in such problems as the loss of surface course stability, rutting, binder bleeding, and fat spots.
- Filler grains bigger than the binder film on aggregates behaves like a filling aggregate, forming mastic, and taking part in filling up the voids among chippings.
- An excess of filler leads to mastic stiffening and the increase of cracking susceptibility.
- The affinity between filler and binder influences the durability of the mix (i.e., its sensitivity to water).
- The appropriate ratios of binder and filler, combined with their properties, have an influence on an SMA mixture's workability and, continuing from that, influence the SMA compaction (or final field density).

One could say that filler is the most underestimated component of SMA. After all, it constitutes from 8% to 12% of a mixture, which actually is a significant amount.

Next we will turn our attention to two concepts used in explaining the behavior of filler—the specific area of a filler and the content of voids in a compacted filler. These concepts can help describe various phenomena occurring in mixes.

3.2.1 Concept of Specific Area

Let us start with the definition of an aggregate's specific area. Specific area is defined as the grains' surface area related to a unit mass, usually given in terms of square centimeter per gram (cm²/g). Fillers are tested in Europe with Blaine's method* according to EN 196-6 (see Chapter 8). The measured specific area depends on how much the parent material was reduced in size (broken up or crushed). The more the material was milled (crushed), the finer the filler, and the larger its specific area.

An example will serve to illustrate the influence of gradation on specific area size. Let us take two fillers and call them A and B. Let them both pass completely (100%) through a 0.063-mm sieve. Laser analyzer tests reveal significant differences in the material smaller than 0.063 mm. There are 20% and 90% of grains smaller than 0.005 mm in fillers A and B, respectively. This means the filler B has a higher content of very fine grains. What difference does this make? A higher content of finer grains means a substantial increase in their specific area. Because of these large differences in the gradation of the fillers that are less than 0.063 mm, the specific areas of fillers A and B may differ widely—even by two or three times—though, at first sight, a simple screening analysis through the 0.063-mm sieve may not suggest such a distinction.

If a filler makes up 10% of a typical SMA mix (100 kg per 1 metric ton), then the specific area of this ingredient alone amounts to 350 million cm² or 3500 m², which is equivalent to the area of a sports field that is 100 m long and 35 m wide. That is quite a lot, isn't it?

In an SMA mixture, the grains' surface area should be evenly covered with a binder film (a couple of microns thick or a bit more). Different specific areas of filler require different quantities of binder to coat them, which means a changeable demand for binder in a mortar. If the true sieve analysis of a filler passes unnoticed and only a simplified screening through the limit sieve is observed (as with fillers A and B), one should not wonder why the same SMA mixture behaves utterly differently when fillers are changed. The specific area that needs to be coated with binder may be substantially different in one case versus another, and a correction of the added binder content may be necessary.

An increase in the specific area of filler requires an increase in the binder quantity just to preserve a suitable mortar consistency. However, an increase in the filler content with a constant binder content causes a reduction of the film-thickness coating the filler grains, making the mix appear to be drier. A substantial increase in the filler–binder (F:B) index increases the risk of cracking.

---

* Blaine's test is not very accurate for it does not take into account fine voids (area textures); more precise measurements can be performed with laser devices (Grabowski and Wilamowicz, 2001).

We can see that the concept of binder content in a mixture based on the specific area size is pretty vivid and appealing to the imagination. There are some analytic formulae approximating the required binder content related to the specific area of a mineral mix; one can find an example in *The Asphalt Handbook* (MS-4).

Before we leave the issue of filler gradation and specific area, it should be noted that much research has raised questions about a direct relationship existing between the gradation and the stiffening properties of a filler (e.g., Anderson, 1987). Also, the significance of the amount of material passing the 0.02-mm sieve and its specific area are questioned as they do not appear to influence mortar properties (Brown and Cooley, 1999). Thus it is time to proceed to another concept, that of voids in a dry-compacted filler.

3.2.2 Idea of Voids in Dry-Compacted Filler

Let us imagine a set of grains that are going to be dry-compacted by tamping.* The result will be a mixture with its volume consisting of grains and some free spaces among them. In a regular binder mortar (blend of filler and binder), these free spaces in a compacted filler would be occupied by binder. The rest of binder would remain as excess filler. Thus binder contained in a mortar can be divided into the following two types (Figure 3.2):

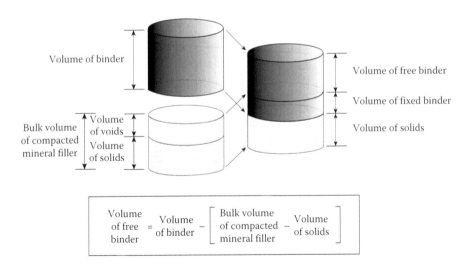

FIGURE 3.2 Free and fixed binder concept. (From Harris, B.M. and Stuart, K.D., *Journal of the Association of Asphalt Paving Technologists*, 64, 54–95, 1995; Kandhal, P.S., *Journal of the Association of Asphalt Paving Technologists*, 50, 150–210, 1981. With permission.)

---

* In this example, the content of voids in dry-compacted filler is being tested after Rigden's method or Rigden's method modified by Anderson (see Chapter 8).

- Fixed binder—binder inside the voids (filling the voids among compacted filler grains)
- Free binder—excess binder remaining after the voids have been filled

As in the comparison of two fillers with differing gradations, here we may demonstrate much the same tendencies—the same quantity of two fillers but with different contents of voids may bond differently to the amount of binder. Actually what really matters is the quantity of free binder, because the properties of the mastic are dependent on it. The lower the content of free binder in a mortar, the faster the growth of its stiffness (Harris and Stuart, 1995). The minimum amount of free binder has been defined in U.S. research, based on the modified Rigden method, as 30% (v/v) of an asphalt mortar (Anderson, 1987; Chen and Pen, 1998). With that level of free binder, filler grains are suspended in the binder and they do not touch each other. In addition, the rest of the mineral mixture (the coarse aggregate) will be coated by only the free binder, so it is important that a sufficient quantity is available.

Figure 3.3 illustrates the process of gradually filling the voids in a compacted filler; the binder essentially plays two roles—that of a lubricant making the relocation of grains easier and that of a liquid in which they can be suspended.*

With a constant content of binder in an asphalt mixture, the quantity of free binder depends on the voids in the compacted filler. With fixed proportions of components in an asphalt mix, the quantity of free binder can be increased by changing to a filler

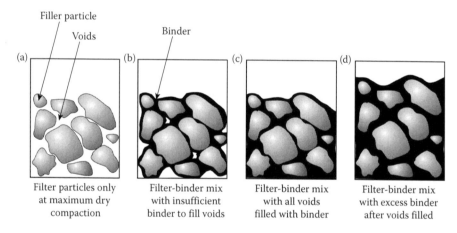

Filler particle

Voids

Binder

(a)   (b)   (c)   (d)

| Filter particles only at maximum dry compaction | Filter-binder mix with insufficient binder to fill voids | Filter-binder mix with all voids filled with binder | Filter-binder mix with excess binder after voids filled |

FIGURE 3.3 Gradually filling the voids in a compacted filler. (From Anderson, D.A., Guidelines on the use of baghouse fines. National Asphalt Pavement Association. Information Series 101–111, 1987b; Harris, B.M. and Stuart, K.D. *Journal of the Association of Asphalt Paving Technologists,* 64, 54–95, 1995. With permission.)

---

* That state might be called a colloidal system of grains (solid bodies) suspended in binder (fluid body). A blend of binder and filler may be regarded as something in between a gel and an alloy at the working temperature of a road pavement (Harris and Stuart, 1995).

with a lower void content. Obviously the reverse is also true. The final requirements for the void content in a compacted filler may be defined as follows:

• They cannot be too high—so as to prevent fixing the whole binder, or too much of it, and to leave enough binder for the rest of the asphalt mix— otherwise the mortar will be too dry, stiff, and susceptible to cracking and water damage
• They cannot be too low, because too much unbonded, excess binder will create a greater risk of mix instability, excessive bleeding and binder drain-down, and the deformability (rutting susceptibility) of an asphalt mixture

The following are the recommended contents of voids in a dry-compacted filler:

• When using results of Rigden's method (test according to EN 1097-4), the mini-mum content of voids in a dry-compacted filler should amount to 28% (v/v) and the maximum content should not exceed 45% (v/v) (Schellenberger, 2002).
• When using results of Rigden's method modified by Anderson (as in the United States) (test according to Anderson [1987]), the maximum content of voids in a dry-compacted filler should not exceed 50% (v/v) (Brown and Cooley, 1999).

The main factors that influence the content of voids in a dry-compacted filler are as follows (Kandhal, 1981):

• Particle size
• Particle shape
• Particle surface structure
• Particle size distribution

Examples of void contents in various dry-compacted fillers tested according to Rigden's original method and specific areas according to Blaine's method [Schellenberger, 2002] are as follows:

Limestone (added filler)	27.7–31.6%	4750 cm^2/g
Diabase (baghouse fines)	30.4–34.2%	3600 cm^2/g
Limestone (baghouse fines)	28.3–32.1%	2280 cm^2/g
Dolomite (added filler)	27.1–28.1%	2068 cm^2/g
Dolerite-microdiabase (baghouse fines)	32.4–36.4%	2658 cm^2/g
Greywacke (baghouse fines)	27.6–31.8%	4054 cm^2/g

The following are some sample results of void contents in various dry-compacted fillers according to Rigden's method modified by Anderson (Schroer, 2006):

Mineral filler	39–47%
Baghouse fines	30–60%
Hydrated lime	66–71%
Fly-ash	37–57%

Generally, the results obtained from Rigden's original method (as used in Europe) are slightly lower than the results from the modified Rigden method (as used in the United States). The reason is the difference in compactive effort. A similar test, which is an indirect usage of the Rigden voids concept, is the German Filler Test; this test shows good correlation to the modified Rigden method results (see Chapter 8).

3.2.3 Mortar—F:B Index

One popular approach used during design practice in many countries is to indicate the recommended range of the filler–bitumen ratio (or F:B index) by weight or volume. Researchers in the United States have said that this factor better describes the maximum content of filler in the mix than does setting specific limits on the filler content. It is also worth adding that those studies have defined the maximum F:B index for asphalt concrete at the level of 1.2–1.5 (by weight) (Anderson, 1987). The F:B index was later altered to 0.6–1.6 in the Superpave method (Superpave Mixture Design Guide. WesTrack Forensic Team Consensus Report, 2001). Finally the suggested F:B ratio for SMA mixes is at 1.5 by weight, taking the total amount of dust on aggregates and added filler as the filler content (Harris and Stuart, 1995). But the F:B index has been criticized for some time, and there are suggestions regarding its replacement by other factors based on the free binder concept. As an example, Australian research studies (Bryant, 2006) suggest the application of an additional filler fixing factor (FFF) apart from the F:B index. Tests have proved that FFF may be also used to estimate the workability of a mixture.

It is necessary to remember that fillers differ markedly in terms of gradation, density, and void content, therefore formulating a universal F:B index is only an approximation. In fact, such an index should be defined for each filler individually. The goal is clear for each case—to produce a mastic that is neither too dry nor too soft.

Figure 3.4 illustrates the relationship between the content of voids in fillers according to Rigden's method and the amount of filler needed to fix binder completely. The higher the F:B index, the more filler is needed to fix the binder. This relationship is illustrated by the graph; the estimated line of completely filled voids in the filler represents the zero-amount of free binder—all the binder is fixed.

As we can see, to fix the binder we need roughly two times more filler that has approximately 30% free voids (the lowest content according to Rigden) than filler that has 50% voids. In the latter case, less filler is sufficient to accommodate all the binder in the free voids (there is a lot of free space for binder within the high void content). It would be difficult for us to use such dependencies unless the filler manufacturer supplies data on the content of free voids according to Rigden or we conduct suitable tests ourselves. All in all, it is better to perform the tests in our own laboratories since eventually a voids parameter may be applied to the entire filler fraction of a mixture (i.e., including the filler fraction that may be coating the coarse and fine aggregate particles).

Another way to evaluate the properties of mortars and the F:B ratio is the application of the softening point (SP) method. According to a publication from Germany (Schroeder and Kluge, 1992) mortars with SPs between 85°C and 100°C perform

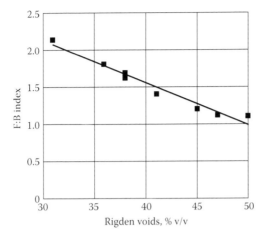

FIGURE 3.4   Dependence of the F:B index on the content of voids in a dry-compacted filler after Rigden's method. (Based on van de Ven, M.F.C., Voskuilen, J.L.M., and Tolman, F. The spatial approach of hot mix asphalt. Proceedings of the 6th RILEM Symposium PTEBM'03. Zurich, 2003. With permission.)

well in SMA. Also the Superpave binder test methods could be used for mortar testing (DSR* and BBR). In Chapter 8, one can find a short description of the methods and some additional remarks.

3.2.4   REVIEW OF MATERIALS APPLIED AS FILLERS

3.2.4.1   Commercially Produced Fillers (Added Fillers)

According to European terminology,† added fillers are made by crushing stone to produce fillers aimed at use in highway engineering. For a long time, the most popular of these has been, and is likely to remain, limestone filler. Limestone filler is distinguished by its affinity with binder, which is one of its strong points. Therefore in Europe limestone filler is most often used for SMA.

The other important feature of industrially manufactured fillers is their repeatability and uniformity of parameters. Finally it is worth observing their constant and repeatable mineralogical composition.

3.2.4.2   Fly Ashes

The use of fly ashes as fillers for SMA is uncommon. Their disadvantages include large specific area (fly ashes are very fine) and the spherical shape of the grains. So fly ashes have only been used to a limited extent and always need an engineering assessment before use.

The density of fly ashes is lower than that of crushed rocks or baghouse fines and fluctuates between 2.0 and 2.6 g/cm³. To obtain a similar volume share in a mineral

---

* DSR - Dynamic Shear Rheometer.
† According to the EN 13043 standard.

blend, ashes are metered in at a lower weight than a standard filler. The modified Rigden void content is usually less than 50% (Report FHWA-IF-03-019, 2003).

3.2.4.3   Hydrated Lime

Many accessible publications (Iwański, 2003; Judycki and Jaskuła, 1999, Little and Epps, 2001) on the use of hydrated lime as an additive to asphalt mixtures have pointed out its positive effect. Apart from a substantial increase in water and frost resistance arising from an improvement in the binder adhesion to the aggregate, an increase in resistance to permanent deformation may also be noticed.

Hydrated lime is distinguished by its strong mixture-stiffening properties at high temperatures. It should be kept in mind that, according to Rigden, voids in hydrated lime can be very large. In connection with that, one should not exceed the standard content of lime in an asphalt mixture, usually accepted as 1.0–1.5% of the aggregate mass. Such an addition of lime will benefit the mixture without the risk of overstiffening.

According to EN 13034, ready-made filler containing hydrated lime is called a *mixed filler.*

3.2.4.4   Baghouse Fines from Asphalt Plants

In some countries a substantial amount of research has been dedicated to baghouse fines, considering them as potential material for use in mixtures. Their use has an economical aspect since the high efficiency of dust collectors in up-to-date asphalt plants collects considerable amounts of dust that are then available for use essentially free of charge. Using some of the dust from dust removal is an everyday practice in many countries. It has been applied to SMA, along with other types of fillers.

The results of studies on the practicality of using baghouse fines are quite divergent. One might conclude that the appropriateness of their use cannot be generalized. Properties of extracted dusts may differ widely, depending on their origin (i.e., source rock type). It is common knowledge that large amounts of collected dust added to mixtures may substantially stiffen them, making them susceptible to cracking and water damage, and asphalt mixes that contain them are not easy to place and compact. Despite that, baghouse fines can be a very good filler (Asphalt Review, December 2004).

Finally it is worth noting one more application aspect of collected baghouse dust. If we want to use baghouse fines as filler, they have to satisfy the standard requirements for fillers, including those concerning the repeatability and uniformity of obtained results. Frequent changes of aggregate types (i.e., rock types) may result in fluctuations in the mineralogical composition of the collected dust and its properties.

The European standard EN 13043 accepts baghouse fines as filler aggregate if they meet the standard requirements for fillers.

3.2.5   Fillers: Résumé

Summing up:

- When designing an SMA mixture, one should not decrease the content of filler below a minimum value (defined by using a 0.063-mm sieve in

Europe or a 0.075-mm sieve in the United States). The lack of filler will be reflected in a decrease in the durability of the mix and the void content and an increase in the risk of the appearance of fat spots.

- An increased amount of filler causes higher viscosity of the mortar, which promotes resistance to deformation. But one should not use an excessive amount of filler because it is easy to overstiffen the mortar, which can lead to cracking.
- Too fine a filler can cause problems since it absorbs much of the binder and may plasticize the mix.
- It is worthwhile to take note of Rigden's test results, particularly when changing filler in the same mixture during continuous production. A new filler with a decidedly different void content than the former one may result in unwanted surprises.
- Adding hydrated lime at a rate of about 1.0–1.5% is a good move as doing so improves the binder-aggregate adhesion and boosts resistance to water damage.

3.3 BINDER

In this section, we will examine the different types of bituminous binders used in SMAs and methods for selecting one.

3.3.1 Types of Applied Binders

Various SMA binders may be seen in the highway engineering practices of many countries. These binders can be divided into paving grade bitumens (unmodified), polymer-modified bitumens (PMB), and special binders (multigrade and others).*

Paving grade bitumens are frequently used. In Europe the most commonly used binder is the penetration graded 50/70 type, and to a lesser extent the 70/100 type. Performance graded binders are routinely used in the United States and may or may not be polymer modified, depending on the base asphalt, the desired range of temperatures at which the binder is expected to perform, and the anticipated traffic level.

Polymer-modified bitumens are increasingly being used. They are found mostly in mixtures laid on roads with high traffic loadings, in special conditions (e.g., road crossings, slow traffic lanes) or on special pavements, although one should remember that PMBs require a suitable technological regime. U.S. experts recommend that highly modified PMBs with polymer contents over 5% (m/m) should not be used (Asphalt Review, December 2004). High polymer contents create problems with fast stiffening and increased difficulties during compaction. Use of highly-modified PMBs combined with other stiffness enhancers should be especially avoided.

---

* According to the EN 12597 standard *Bitumens and binder products. Terminology,* "hot" applied bitumens may be divided into road bitumens (soft and hard), modified (including polymer modified) and special.

Properly chosen and tested modified binders could increase rutting resistance and decrease the risk of low-temperature cracking of an SMA pavement.

Among the special binders, multigrade ones are sometimes used in SMA mixtures; for example, in Australia (NAS AAPA, 2004) they are used for heavily trafficked pavements.

3.3.2 SELECTION OF A BINDER

In countries where paving grade bitumens (unmodified) are used in SMA, it is usually assumed that the application of hard binders to ensure improvement in rut resistance is not necessary. It is generally accepted that, ensuring rut resistance should be accomplished by creating the correct mineral skeleton. So medium-grade binders are justified for use in SMAs, such as the popular 50/70 binder used in Germany or the 70/100 used in the Netherlands. A very soft binder like a 160/220 may be seen in Europe (e.g., in Finland); in very cold climate conditions, use of such a soft binder in not surprising.

The selection of a binder for an SMA wearing course is determined, on the one hand, by the temperature range over which the pavement is expected to perform and, on the other, by the expected traffic loads. The type of binder selected is limited by the local temperature conditions, which in most cases disallow for the application of excessively hard binders to prevent low-temperature cracking. In cold climate countries, softer PMBs are applied due to their high elastic recovery at low temperature. Some examples of these binders are presented in Chapter 5. A tendency to change the type of binder in consideration of the increase in traffic load—from paving grade, through multigrade, and up to a low penetration PMB—is clearly evident.

Despite particular emphasis being placed on securing rutting resistance mainly by the SMA skeleton, there is no doubt that the proper selection of a binder is an extra element supporting the stone skeleton performance. German research studies (Graf, 2006; Kreide, 2000) show that in most cases modified binder significantly increases SMA rutting resistance.

3.4 SUMMARY

- SMA mastic consists of fine (or passive) aggregate, filler, stabilizer, and bituminous binder. Binder and filler together create mortar.
- One reason for using fine aggregate is to fill the voids among the coarse (active) grains and participate in their interlocking.
- The term *filler* denotes all the aggregate that passes through the limit sieve (0.063 mm in Europe or 0.075 mm in the United States). It is suggested that the properties of the whole filler fraction passing through limit sieve (i.e., added filler plus aggregate fines) be tested. Test results for the entire filler fraction may differ from the results of testing only the added fillers.
- An excess of filler leads to the stiffening of the mortar or mastic and an increase in its susceptibility to cracking.
- The correct filler–binder (F:B) ratios and mortar properties have an impact on the workability, compaction, and permeability of a given mixture. In

fact, formulating a universal F:B index is a kind of approximation; such an index should be defined for each filler individually.

- Fillers with high contents of voids (analyzed according to any Rigden's method) should be used cautiously due to the insufficient content of free binder in a mortar, which results in stiffening the mortar.
- Adding hydrated lime at a rate of about 1.0–1.5% (m/m) is a positive solution, improving water resistance and the adhesion of the binder-filler to the aggregate.
- SMA binders can be divided into paving grade (unmodified) binder, PMB and special (most often multigrade) binder. Each of these types can be successfully used, providing the mix is well designed with properties appropriate for the traffic loading and local climate.
- The application of hard (low-penetration) paving grade binder (e.g., 35/50) for wearing courses is not recommended due to the high risk of low-temperature cracking.

4 Stabilizers (Drainage Inhibitors)

Stone mastic asphalt (SMA) mixtures require a high content of binder, which results in thick binder films on the aggregate grains. To avoid the draindown effect, stabilizing additives (drainage inhibitors) are indispensable in most cases. This chapter describes the types of stabilizers and methods of testing them.

4.1 THE DRAINDOWN EFFECT

Have you ever seen fat spots on an SMA surface? Or binder running out of a truck hauling a hot SMA mixture? If you have, those troubles may have been caused by a binder or mastic draindown.

An SMA asphalt mixture has an intentional binder surplus. The specific surface of the mineral mixture is too small in relation to the designed binder volume. Under normal conditions, that binder is not bonded with the mineral mixture grains and does not remain on the grains' surface; instead it drains-off. The draindown effect results from the separation of part (binder or mastic) of the SMA asphalt mixture. Keeping in mind that SMA has a lot of binder—in fact, SMA has a deliberate excess of binder—one should always take into account the risk of draindown.

In many countries, in the early applications of SMAs there were some cases of hot binder running out of a silo that held a hot SMA mix. Similar occurrences took place out of the backs of trucks carrying SMA to construction sites, eventually appearing as fat spots (bleeding) on the finished surfaces.

Methods to determine the amount of draindown are explained in Chapter 8 and other problems that may occur on the construction site in Chapter 11.

4.2 STABILIZERS (DRAINAGE INHIBITORS)

In the 1960s—during the beginning stages of SMA manufacture—the need arose to incorporate stabilizing agents to prevent binder draindown. Such additives are called stabilizers; they stabilize or keep the binder in place. Because of these stablizers, an increase in the binder film thickness on the aggregate is possible.

The two main techniques of reducing binder draindown are as follows:

- Additives that absorb part of the binder (the surplus that is likely to draindown)
- Additives (polymers) that increase binder viscosity at high temperatures, which in turn reduce the risk of its draindown

Each of these methods will be discussed later in this chapter, but let us start with a definition of a stabilizer. A *stabilizer* or *drainage inhibitor* is an additive to an SMA asphalt mixture put in to prevent binder (or mastic) from draining-off. Stabilizers may be made up of various materials, including both binder absorbers (e.g., fibers) and viscosity boosters (e.g., polymers).

In addition to helping to retain binder on the aggregate, stabilizers may also improve other properties of the binder itself, the mastic, or the mixture (see Section 4.2.1.4). There are examples of such tests available in the literature—for an example, see Behbahani et al. (2009).

But are stabilizers really necessary for SMA? Research done in the United States during the 1990s revealed a 70 times higher binder draindown in a mix without any stabilizer compared with the same mix containing 0.3% of cellulose fibers (Brown and Mallick, 1994).

The author's experience at the beginning of 1990s showed that even with using modified binders, fat spots often occurred. Then trials testing modified binders with half of the typical amount of stabilizer were conducted, and fat spots also occurred. Now, independent of binder type, at least a minimal amount of stabilizer is commonly used. So the answer to the question, should be stabilizer used or not? is evident; however, one has to note that some compositions of SMA are less susceptible to draindown.

4.2.1 Binder-Absorbing Additives

Binder-absorbing additives are the most popular SMA stabilizing agents. The following properties are required of a stabilizing material:

- Adequate binder absorbing power—this is the most significant property
- Ability to act without weakening the mixture—the stabilizer must not create glide planes and lessen the grain-interlocking strength.

Stabilizers of this type occur in various forms related to the following kinds of raw materials:

- Cellulose—the most popular (Shown in Figure 4.1)
- Pseudocellulose—made of milled or fragmented waste paper
- Mineral fiber—developed through melting rocks and subsequently processing the melted rock to form threads (like rock wool)
- Cellulose–mineral—a blend of cellulose and mineral fibers occurring in various compositions (proportions)
- Cellulose–polymer—a blend of cellulose fibers and different types of polymers occurring in various compositions (proportions)
- Cellulose–wax—a blend of cellulose fibers and synthetic waxes, which not only stabilizes but changes the binder viscosity–temperature relationship as well
- Textile—threads of processed and fragmented textile waste products
- Plastics—for example, polypropylene (Shown in Figure 4.2)

- Glass—in the form of threads (like fiberglass wool)
- Others—for example, leather waste products (leather dusts)

The crucial difference among stabilizers is their absorbing power. To date, the most effective of the binder-absorbing stabilizers are cellulose fibers. They have a very high-binder absorption, which results in holding the binder firmly in position.

Fiber-free stabilizers have a wide range of binder absorption powers, so any new product should be tested in the laboratory every time (see Chapter 8). One should also remember that stabilizers may have substantially different densities, which are directly translated into their volumes in a mixture. High-density mineral fibers need a higher addition rate at batching, usually on the order of 0.4–0.6% (m/m). The same is true of polypropylene and glass fibers.

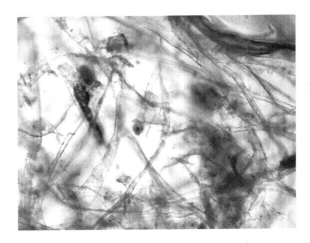

FIGURE 4.1 Cellulose fibers (threads)—microscope image, × 100 enlargement. (Photo courtesy of Jan B. Król, Warsaw Technical University.)

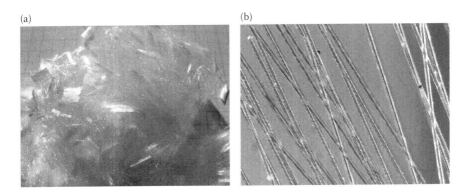

FIGURE 4.2 An example of 6-mm long polypropylene fibers: (a) macroscopic view, (b) the same fibers in microscope image, × 100 magnification. (Photo [a] courtesy of Krzysztof Błażejowski and photo [b] courtesy of Jan B. Król, Warsaw University of Technology.)

FIGURE 4.3 Plastic fibers can be an SMA stabilizer but must be used at a high-addition rate. The photo shows an SMA mix with 0.6 % (m/m) of polypropylene fibers. (Photo courtesy of Halina Sarlińska.)

Furthermore, if a stabilizer does not work efficiently enough, increasing its quantity may improve the draindown performance. However, larger amounts of stabilizers may cause unexpected troubles, such as decreasing mixture workability. For example, plastic fibers (Figure 4.3) are marked by a lower absorption power. Using a high quantity of them (say, about two times more than cellulose fibers) will produce a substantial growth of stiffness and low workability. Small wonder—6 kg of such fibers per 1 metric ton of an SMA translates into a substantial volume; fibers are clearly visible in the mix as shown in Figure 4.3.

The most popular stabilizers—namely, products of cellulose—occur in the following forms:

- Loose fibers—in the form of irregularly shaped cellulose threads
- Pellets—granulated products without a binding agent
- Granulated products—cylinder-shaped granules that consist of threads coated with bituminous binder or another agent (e.g., wax or plastic)

4.2.1.1 Loose Cellulose Fibers

Loose cellulose threads (Figure 4.4) have the longest history of application. One of their advantages is that they become effective immediately after mixing them with aggregate and binder. They should be protected from moisture as they are highly hydrophilic and easily absorb water. Wet fibers are not able to absorb the binder and therefore are not effective.

Loose cellulose fibers are supplied in bags of thermo-shrinkable plastic. The mass of fiber per bag can be prearranged with the manufacturer and should be tailored to the batch volume of a given batch from an asphalt-mixing plant. An automatic

FIGURE 4.4 An example of loose cellulose fibers. (Photo courtesy of J. Rettenmaier & Söhne GmbH + Co. KG, Germany.)

metering process for SMA production in an asphalt-mixing plant has been possible for some time now; loose fibers are delivered by autotankers and stored in a silo. During mix production, they are blown into the pugmill via an automatic system. The same pneumatic metering system can be used in drum-mixing plants.

When adding loose fibers, it is worthwhile to remember the following:

- The time at which the fibers are batched into the mixer is very important; bags should be emptied into the mixer during the dry mixing time, prior to binder loading.
- An excessive increase in the dry (binder-free) mixing time may lead to grinding of the fibers into dust by the aggregate, limiting the effectiveness of the stabilizing action.
- Loose fibers are very sensitive to moisture and therefore should be properly protected and stored; wet fibers lose their absorbing power.

When testing SMA mixtures in a laboratory, loose fibers may be treated without any special precautions. A mechanical mixer is not required.

4.2.1.2 Pelletized Loose Fibers

Loose fibers may be pressed into pellets to keep the fibers together without binding agents. Their shape makes metering of the SMA mix during production easier. Pelletized fibers are formed into different shapes; one example is presented in Figure 4.5. Automatic loading with pneumatic feeders is also possible and easy. The pellets are usually supplied in large packages (big-bags) and stored in silos. All other features of loose fibers remain unchanged.

When performing tests of SMA mixtures in a laboratory, pelletized fibers may be treated without any special preparations. A mechanical mixer is usually not required;

FIGURE 4.5   An example of pellets—loose formed fibers. (Photo courtesy of J. Rettenmaier & Söhne GmbH + Co. KG, Germany.)

FIGURE 4.6   Examples of granulated fibers (a) made of waste paper with wax coating (Photo courtesy of Excel Industries Ltd., U.K.) and (b) made of cellulose with binder coating. (Photo courtesy of J. Rettenmaier and Söhne GmbH + Co. KG, Germany.)

however, in the case of very dense pellets, it is better to use one to ensure that the pellets are broken up and thoroughly mixed.

4.2.1.3   Granulated Fibers

Granulated products (Figure 4.6a and b) are produced by coating fibers with a binder or other binder-soluble agents. A binder coating enables control of the forces on fibers during granulation and separation of individual fibers, which is necessary for distributing them evenly in an SMA mix.

The granulated form of these products makes dosage at production easy. The fiber granules are supplied in big-bags or autotankers and are stored in the bags or silos. Automatic loading with screw-pneumatic feeders is also possible.

Special attention should be paid during the preparation of laboratory samples. Due to compression of the granules during production, a higher shear force is needed to distribute the fibers; therefore one should use a mechanical mixer. If samples must be mixed by hand, it is recommended to warm the granulated product in an oven to a temperature above the softening point of the coating agent. Otherwise the stabilizer will not work effectively in laboratory samples, possibly leading to unrealistically high-binder draindown.

Granulated products are the best options for SMA production at medium and large-capacity asphalt plants because of the ease of handling.

4.2.1.4 Granulated Cellulose Fibers with Additives

Several mixtures of cellulose fibers and various modifiers are also available. These types of compounds have been produced with two goals in mind—binder stabilization and a change in the specific characteristics of a mix (e.g., workability or durability). These compositions enable the simultaneous feeding of mixtures with fibers and additive. The effectiveness of such products should be tested in the laboratory to confirm the final properties of the asphalt mixture.

4.2.1.5 Packing and Delivery Forms

Loose fibers are packed in self-shrinkable plastic bags. They are thrown into a mixer intact and the film melts into the binder during mixing. Granulated cellulose fibers may be bagged like other fibers (Figure 4.7) or packed in big-bags (Figure 4.8). Both loose and granulated fibers can be supplied in autotankers (Figure 4.9).

FIGURE 4.7   Granulated stabilizer in PE bags of various mass—from 3 to 10 kg. (Photo courtesy of J. Rettenmaier and Söhne GmbH + Co. KG, Germany.)

FIGURE 4.8 Dosing method of granulates—emptying a big-bag onto a conveyor belt. (Photo courtesy of Excel Industries Ltd., U.K.)

FIGURE 4.9 Tractor-storage bin unit loaded with granulated stabilizer. (Photo courtesy of Excel Industries Ltd., U.K.)

4.2.1.6 Summary

The following is a comparison of loose or pelletized fibers and granulated fibers:

- Loose or pelletized fibers, packed in bags of a fixed weight, are easy to throw manually into a pugmill at a batch asphalt plant, though the method is problematic for big contracts, high-output plants, and plants without semiautomatic metering. Loose fibers have their staunch fans, claiming that this type of fiber can be better distributed in a mix and that a lesser amount of them (compared with granulated products) is needed to achieve a similar effect. These fibers do not present any problems during laboratory testing, and they require no extra treatment or equipment. Their disadvantages lie in water absorption, which requires covered storage, and the necessity of manual operations at batching.
- Granulated fibers usually contain the same fibers as loose fibers, except that they have been coated with a binder (or other material), enabling production of granulated products that are 3–4 mm in diameter and 3–5 mm long. The binder coating makes them more water resistant (i.e., less moisture-sensitive). Another advantage of granulated fibers is the option of automatic dosage into the mixer at an asphalt plant, usually through a screw-pneumatic feeder. Granulated fibers do require a suitable temperature and mechanical mixing to release them and allow them to be dispersed properly in the SMA during production, so granulated products must be added at the right moment and mixed properly.
- Another significant issue that should be considered is the susceptibility of fibers to water. It is little wonder that a stabilizer absorbs water if it is meant to absorb binder. A wet stabilizer may harm the mixture considerably (see Chapter 11). Loose fibers or pellets are more sensitive to moisture than coated granulated products.
- Various combinations of fiber blends have been appearing here and there (e.g., cellulose–mineral and cellulose–polymer). Some of them reveal certain additional features, such as building up the mastic and enhancing its shear strength.
- The amount of fibers in a mix are determined using Schellenberg's method or other methods (see Section 4.5 and Chapter 8), normally at a level of 0.3–0.4% (m/m).

4.2.2 BINDER VISCOSITY PROMOTERS

The second way to counteract the draindown phenomena is to use a binder viscosity promoter. The most common agents used to be polymers, either plastomers or elastomers. Test results show, however, that the efficiency of polymers is not as good as that of fibers. Currently in many countries, regardless of the type of applied binder (modified or paving grade), fiber stabilizers are required. Binder modified with special materials may be considered, except for polymers directly metered into a pugmill at an asphalt plant. Polymer stabilizers also have a higher viscosity over the preferred range of production temperatures. This is when the use of a classic stabilizer is indispensable because, as a number of failed attempts have shown, modified binder itself does not protect an SMA mix from segregation.

The application of viscosity promoters acting at temperatures of 100–160°C may also increase problems with compaction on site because an SMA mix is hard to compact, even without an additive that stiffens the binder and makes compacting the mix even more difficult. This fact should not be forgotten when choosing a stabilizer to use.

4.3 PROPER SELECTION OF THE AMOUNT OF A STABILIZER IN A MIX

The procedure for the quantitative selection of a stabilizer in a mix is simple, although it may be a little laborious. One of the draindown testing methods described in Chapter 8 may be adopted here. A series of SMA samples of the same composition should be prepared, but with different quantities of stabilizer. If the procedure is carried out with a proven product (e.g., cellulose fibers), testing may be limited to 0.2%, 0.3%, and 0.4% (m/m). If it is a product that has not yet been used, the testing series ought to cover a somewhat wider range based on the expected behavior (e.g., 0.1%, 0.2%, 0.3%, 0.4%, and 0.5% [m/m]). Draindown should be determined for each stabilizer content, and the results should be plotted as shown in Figure 4.10. An important remark—the test temperature should be as close as possible to the real production temperature of the mix in the asphalt plant (see Chapter 8). The amount of stabilizer required for complying with the contract specifications may then be read off the diagram.

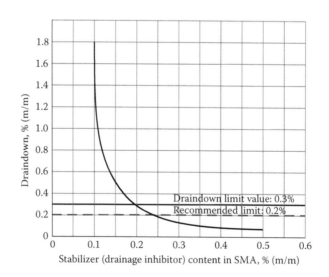

FIGURE 4.10 Example relationship between draindown and stabilizer content used for the selection of an effective content of stabilizer in an SMA mix.

The optimal content of a stabilizer is the one that produces less than 0.3% by mass draindown material (recommended value is less than or equal to 0.2% m/m), unless otherwise specified.

4.4   TESTING METHODS FOR STABILIZERS

Despite the great impact of the stabilizer on SMA quality, few countries have developed testing methods and criteria for the practicality of stabilizers. Some assessment methods for the application of stabilizers in various countries are presented in Chapter 5.

Practice proves that a reliable stabilizer-effectiveness assessment can be carried out when the same SMA mixture (the same JMF[*]) is tested with different stabilizers. Comparison of test results from tests on different SMA mixtures is unreliable. The results of one test will not be comparable with the those of the others because the results for each mixture depend on the SMA composition, such as the type of aggregate and binder contents.

A direct method of checking the effectiveness of stabilization (i.e., absorption capacity) is the oil-absorption method (see Chapter 5).

4.4.1   STABILIZER TEST RESULTS

Almost every working road laboratory has conducted many series of draindown tests. The results of companion tests on a wide variety of stabilizers are seldom published. However, the following interesting conclusions can be drawn from large-scale research conducted in the United States (Brown and Mallick, 1994):

- SMA binder draindown was seen to be a time-progressive process; the fastest run-off occurred within the first 30 minutes of hot mix storage, then it continued over 2 hours during testing. Therefore one can conclude that the major part of the draindown process takes place during the 1st hour after SMA production.
- The most effective stabilizers tested were cellulose and mineral fibers.
- Differences between binder-absorbing stabilizers and viscosity-enhancing stablizers were observed.
- Lack of a stabilizer caused as much as 4% (m/m) mastic drain-off in a 0/19 mm coarse-graded SMA.

During the course of a season of roadwork, we decide, for various reasons, to change the supplier of stabilizers, but we do not always have enough time to check the SMA draindown with the new product (after all, it looks more or less the same). At that time, we have to take into account some unexpected problems and be prepared to react quickly—for example, to increase the quantity of stabilizer, if needed.

---

[*] JMF - Job Mix Formula - recipe of the mixture.

4.5   BINDER DRAINDOWN TESTING

Various methods for testing the tendency for binder to draindown in a particular SMA are presented in Chapter 8. The original Schellenberg's method and other techniques according to AASHTO and EN 12697-18 standards are also described there.

In most countries, Schellenberg's or a similar method for determining the quantity of stabilizer in a mix has been applied. The maximum allowable mass of drained-off material is essentially the same almost everywhere and is equal to 0.3%. A recommended, safer limit is 0.2%.

4.6   SUMMARY

Factors influencing a mixture's susceptibility to binder draindown are presented in Table 4.1.

TABLE 4.1
Summary of Factors Influencing Draindown Potential in SMA

	Effect on Draindown Potential	
Factor	By Decreasing Factor	By Increasing Factor
Stabilizer content	Increase	Decrease
Stabilizer Type		
• Binder absorbers	Strong-to-medium decrease	
• Viscosity boosters	Medium-to-weak decrease (depending on type and temperature characteristics of used polymer)	
SMA Mixture Composition		
• Effective content of binder	Decrease	Increase
• Content of sand fraction	Increase	Decrease
• Content of filler	Increase	Decrease
Components' Properties		
• Rigden voids (voids in a dry compacted filler)	Increase	Decrease
• Aggregate absorption power (inversely related to effective binder content)	Increase	Decrease
Mixture Production Parameters		
• Production temperature	Decrease	Increase
• Silo storage time of a hot mixture	Decrease	Increase

Note: SMA = stone mastic asphalt.

From the point of view of an SMA mix producer, stabilization efficiency is a decisive factor. But there is one more, equally important point: the quality of a stabilizer must be constant over the whole season of use. What does it matter, exactly, if a product is best at the moment of testing? If the quality varies, we may find ourselves wondering, "What happened? This cellulose is not as good as it used to be." This variable quality often results in fat spots on the compacted SMA surface.

5 Requirements for SMA Materials

A distinctive feature of stone matrix asphalt (SMA) components is their high quality. In this chapter we shall have a look at the requirements for those materials in various countries. Comparing them may be interesting because of the diversity of emphases different countries place on individual conditions.

Consequently, the following components will be subsequently discussed in detail:

- Aggregates
- Binders
- Stabilizers
- Reclaimed asphalt
- Other additives

5.1 REQUIREMENTS FOR AGGREGATES

The requirements for aggregates are quite diverse, and they chiefly concern fundamental properties that influence the performance of SMA and are attributed to the size of the grains. This has given rise to a variety of requirements for particular fractions—some for coarse (active) aggregates and other for fine ones. The requirements for fillers have been defined in Chapters 2 and 3.

One may find that almost everywhere these requirements center around the following properties:

- Coarse aggregate: hardness (resistance to fragmentation or crushing), percentage crushed, shape of grains, polishing resistance, and resistance to external factors (e.g., water, frost, deicers)
- Fine aggregate: angularity, amount of dust, and absence of swelling fractions in dusts
- Fillers: stiffening properties—difference in softening points (SPs), absence of swelling fractions, and air voids in a compacted filler.

Taking into account the type of aggregate skeletons required for SMAs, the primary property seems to be resistance to crushing. This resistance is often assessed using the very popular Los Angeles abrasion test (L.A. index). The range of the allowable LA index in national specifications typically varies from 20 to 30%; however, many publications from North America reported that some successful SMAs have aggregates with LA index values that are greater than 30%.

In one report (Celaya and Haddock, 2006) the authors express their opinion that the LA index is not the only important property of coarse aggregates for SMA. Besides the LA index, one should take into account the Micro-Deval test, which is conducted using water* and the Strategic Highway Research Program (SHRP) Superpave Gyratory Compactor (SGC) compaction degradation. This difference between the LA index and Micro-Deval test can show the aggregates that are sensitive to crushing in the presence of water. Finally, the results from the LA index, Micro-Deval test, and SGC degradation test provide the best way to select suitable coarse aggregates for use in SMA mixtures. Comprehensive reviews of coarse aggregate test methods are available in the published literature such as Fowler et al. (2006) and White et al. (2006).

In Germany, the Schlagzertrümmerungswert (EN 1097-2, Part 6) method is used instead of or together with the LA index. In general, resistance to crushing is crucial and must be tested because the use of weak aggregates may result in poor performance of the whole layer (Figure 5.1).

For the fine aggregate portion, the most commonly specified property is angularity. This can be tested using different methods; common ones include the flow rate method as in European Standard EN 933-6 (results in seconds, shown as Ecs) or the measurement of voids in uncompacted aggregate as in AASHTO T 304 (results in % [v/v], known as fine aggregate angularity [FAA]). Obviously other methods are used and could also be good indicators of angularity. Johnson et al. (2004) have conducted interesting research to evaluate the influence of FAA and

FIGURE 5.1   An example of the use of very weak aggregates in an SMA mixture; sample after wheel tracking test shows fractured aggregate particles. (Photo courtesy of Krzysztof Błażejowski.)

---

* This method, which is described in European Standard EN 1097-1, could also be performed without water. In the United States the Micro-Deval method is based on AASHTO T327 (ASTM D6928).

other variables on asphalt mix performance. That study confirmed that FAA is a good tool to predict dynamic modulus and rut resistance.

The requirements for filler, besides gradation, are focused on analyses of clay and silt in the filler and on the stiffening properties represented by an increase of the SP and Rigden voids. The specific areas of filler are gradually being withdrawn from specifications in some countries (e.g., Poland and Finland). Applications of different raw materials as fillers were described in Chapter 3. Some waste materials, such as coal fly ash, waste ceramics, and steel slag, have also been tested, and the results indicate a good potential for use in SMA (Muniandy et al., 2009).

The requirements for aggregates in selected countries are described throughout this chapter. The data are divided into two parts—that for CEN countries (mainly European Union) and that for the United States. A more comprehensive record of data would surpass the scope of this book.

5.1.1 Requirements According to European Standard EN 13043

In the Comité Européen de Normalisation, or European Committee for Standardization (CEN), the member states' requirements for aggregates designed for asphalt mixtures have been unified in the EN 13043 standard entitled *Aggregates for Bituminous Mixtures and Surface Treatments for Roads, Airfields, and Other Trafficked Areas*. It provides a set of aggregate properties and a series of requirement levels (categories) for each property.

Each CEN member state adopting this standard has defined its own national requirements, considering such issues as local climatic conditions and experience of engineering, from among the alternative properties and categories provided in the standard.

The requirements for SMA aggregates according to the EN 13043 standard are displayed in Tables 5.1 and 5.2. The following is an explanation of records in the Tables 5.2 through 5.4:

- The notation *Declared* exclusively denotes the necessity of giving a test result without determining any threshold limit for that requirement.
- The notation *Category* refers to the following:
  - An absolute numerical value for a given category in the system presented in EN 13043: the numerical limit is shown in the table following a letter symbol for a given category (see also Table 5.1 for explanations)—for example, a category MB_F10 means the requirement is MB_F is less than or equal to 10 g/kg according to test method EN 933-9 or PSV_{50} means requirement PSV is great than or equal to 50 according to test method EN 1097-8).
  - A declared value, which means the result of the test is outside the bounds of the last category with a specified numerical value—for example, category MB_FDeclared is used when test result is larger than 25 g/kg. For two specified properties: resistance to polishing (PSV) and resistance to abrasion from studded tires (AN), declared values mean both—result of the test outside the limit or any intermediate value.

- No requirement category (NR), which means that in a given country's national specification, the requirement for this property is not used, e.g., $MB_F NR$
- Fractions of aggregates are described as d/D (e.g., 2/5 mm) where:
  - *D* means nominal upper limit of gradation (oversized grains are allowed)
  - *d* means nominal lower limit of gradation (undersized grains are allowed)

Some additional explanations are provided in Table 5.1, but details can be found in EN 13043. An example of a similar type of EN standard is described in Chapter 14.

Tables 5.2 through 5.4 present the requirements from selected European countries. The substantial differences among these countries may be confusing. In the most important properties, the demanded level is more or less similar, e.g., LA index is from 20 to 25% or flat and elongated content is usually as high as 20–25%.

Although it is the same SMA mixture, the requirements for components or aggregates, depend on the following factors:

- Materials available in a specified place—countries specify their requirements based on long-term experience with aggregates, test results, and research. Requirements also depend on accessible sources of aggregate—countries in which sufficient amounts of very good quality materials exist are able to limit the number of required properties. One example of such a situation are the Nordic (Scandinavian) countries. However, several countries have quite a wide range of aggregates with very different qualities. There is a need to balance the technical requirements with a view toward economics.
- Technology and previous experiences with materials also have an impact on the set of requirements. For example, using hydrated lime (in mixed filler) is not very popular in Europe, therefore only a few countries put them into their specifications (filler category *Ka*).
- Test methods in EN 13043 come from different national practices, as described in Chapter 14, hence many countries do not use some of them in practice. For example, the LA method has not been commonly used in Germany, and the same can be said for Rigden's method in Poland, Aggregate Abrasion Value (AAV) outside the United Kingdom, and the Nordic abrasion value outside Scandinavian countries. Many countries use *declared* categories because without past experiences it was hard to establish any reasonable requirement.
- In specified number of countries, production of aggregates for asphalt layers is regulated with precisely prescribed fractions and their individual gradation limits. Such a situation, in which all producers of aggregates make the same fractions and with very similar gradation, is very comfortable for asphalt mix producers. At the same time any additional

TABLE 5.1
Comments on System of Requirements Based on EN 13043

Properties	Label of Category	Comments
Grading, according to EN 933-1	Coarse aggregates: G_CX/Y	Gradation of aggregate fractions after EN 13043 are labeled as G_CX/Y category; the numbers describe allowable amounts of oversized (X) and undersized (Y) material, e.g. $G_C90/15$ means that only 10% of oversized and 15% of undersized material is allowed, similarly $G_C85/15$ means that only 15% of oversized and 15% of undersized material is allowed; category $G_C90/10$ is the highest possible choice and $G_C85/35$ is the lowest.
	Fine aggregates: G_F85	There is only one category for fine aggregate, G_F85, which means the limit for oversized grains (>2 mm) is 15% by mass.
	All-in aggregates: G_AX	When all-in aggregates are used one of two G_AX categories, can be used G_A90 and G_A85, where maximum limits for oversized material are 10% and 15% (by mass), respectively.
Tolerances of typical gradation	Coarse aggregates: $G_{X/Y}$	The idea of this type control is based on a requirement that the producer will document and declare the typical gradation of any produced aggregate fraction; tolerances of typical gradation, labeled as $G_{X/Y}$, are used for gradation control within the fraction; depending on the D/d coefficient, the control sieve is chosen as D/2 or D/1.4; the number x means x—overall limits of amount of material passing by control sieve (20 means 20–70% by mass; 25 means 25–80% by mass), and the number y means y – tolerances for typical gradation on control sieve declared by aggregate's producer (15 means ±15%; 17.5 means ±17.5%); example category $G_{25/15}$ means that on sieve D/2 or D/1.4 overall limits are 25–80% and the tolerance from the declared value is ±15%
	Fine and all-in aggregates $G_{TC}X$	This requirement applies to tolerance of percentage of grains passing by sieves D, D/2, and d compared with the gradation declared by producer; category $G_{TC}10$ mean tolerances on sieves: D ±5%, D/2 ±10%, d ±3%, category $G_{TC}20$ mean tolerances on sieves: D ±5%, D/2 ±20%, d ±3%,

(Continued)

TABLE 5.1 (CONTINUED)
Comments on System of Requirements Based on EN 13043

Properties	Label of Category	Comments
Fines content, according to EN 933-1	f_X	Amount of grains passing by 0.063-mm sieve, where x in f_X category means maximum allowable content of fines
Fines quality, according to EN 933-9 (methylene blue test); in fine and all-in aggregates	MB_FX	Control of harmful fines (e.g., swelling clay); category MB_FX means that maximum X methylene blue value (g/kg) is allowed in fines; used only if fines content is between 3% and 10% by mass of material; if fines content >10%, requirement for filler applies
Angularity of fine aggregates according to EN 933-6, p.8	$E_{cs}X$	Flow coefficient of fine aggregates, labeled with $E_{cs}X$ category, where X means minimum time of flow in seconds
Shape of coarse aggregate (Flakiness Index EN 933-3, Shape Index EN 933-4)	SI_X or FI_X	FI or SI could be used for determination of grains' shape; labeled as SI_X and FI_X, where maximum allowable amount of flat and elongated particles is marked as X (% by mass)
Percentage of crushed and broken surfaces in coarse aggregates, according to EN 933-5	$C_{X/Y}$	Percentage of crushed and broken surfaces labeled as $C_{X/Y}$, where X means percentage of completely broken particles (by mass) and Y means percentage of completely rounded particles (by mass); so a requirement to use only crushed coarse aggregates is described in category $C_{100/0}$. Category $C_{95/1}$ allows up to 1% of noncrushed particles in aggregates.
Resistance to fragmentation (crushing), according to EN 1097-2 clause 5	LA_X	Resistance to crushing (fragmentation) with LA method; labeled as LA_X, where maximum allowable LA coefficient is marked as X (%)
Resistance to fragmentation (crushing), according to EN 1097-2 clause 6	SZ_X	Resistance to crushing (fragmentation) with German Schlagzertrümmerungswert method (impact test); labeled as SZ_X where maximum allowable SZ coefficient is marked as X (%)
Resistance to polishing according to EN 1097-8	PSV_X	PSV; labeled as PSV_X, where X means required minimum PSV value
Resistance to surface abrasion, according to EN 1097-8 Annex A	AAV_X	AAV; labeled as AAV_X, where X means maximum allowable AAV value
Resistance to wear, according to EN 1097-1	$M_{DE}X$	Micro-Deval coefficient; labeled $M_{DE}X$, where X means maximum allowable Micro-Deval value
Resistance to abrasion from studded tires, according to EN 1097-9	A_NX	Nordic abrasion value; labeled as A_NX, where X means maximum allowable value

TABLE 5.1 (CONTINUED)
Comments on System of Requirements Based on EN 13043

Properties	Label of Category	Comments
Water absorption, according to EN 1097-6	$W_{cm}0.5$ $WA_{24}X$	The method of testing is chosen depending upon the size of the aggregate: • Using EN 1097-6 clause 7 refers to category $WA_{24}X$, where X means maximum allowable percentage absorption by mass. • Using EN 1097-6 Annex B refers to category $W_{cm}0.5$, where 0.5 means maximum allowable percentage absorption by mass (there is only one category ≤0.5%). Additionally EN 13043 connects water absorption and resistance to freeze-thaw of aggregates; aggregates with small absorption are assumed to be freeze-thaw resistant.
Resistance to freezing and thawing, according to EN 1367-1	F_X	Category F_X is used, where X means maximum allowable percentage loss of mass; test can be conducted in water, salt solution, or urea.
Resistance to freezing and thawing, according to EN 1367-2	MS_X	Category MS_X is used, where X means maximum allowable percentage loss of mass; test is conducted with magnesium sulfate.
Resistance to thermal shock, according to EN 1367-5:	—	Test of aggregate resistance for high temperature; results are declared
Affinity of coarse aggregates to bituminous binders, according to EN 12697-11	—	Test of binder adhesion to aggregate; results are declared
"Sonnenbrand" of basalt, according to EN 1367-3 and EN 1097-2	SB_{SZ} or SB_{LA}	This is to check for basalt rock decay, which results in lowering aggregate strength and in most cases in very low freeze-thaw resistance of basalt aggregate; categories SB_{SZ} and SB_{LA} mean that after test (boiling for 36 hrs) and crushing (in SZ or LA, respectively), the aggregate must meet required values: • Loss of mass after boiling: max 1.0% (and) • Increase of impact value: max 5.0% (SB_{SZ}) (or) • Increase of LA coefficient: max 8.0% (SB_{LA})
Coarse lightweight contaminators, according to EN 1744-1, p.14.2	$m_{LPC}X$	The content of coarse lightweight organic contaminants larger than 2 mm should be maximum X% by mass.

(Continued)

TABLE 5.1 (CONTINUED)
Comments on System of Requirements Based on EN 13043

Properties	Label of Category	Comments
Dicalcium silicate disintegration of air-cooled blastfurnace slag, according to EN 1744-1, p. 19.1	Resistance required	Slag aggregate will be free from dicalcium silicate disintegration, the results are declared.
Iron disintegration of air-cooled blast furnace slag, according to EN 1744-1, p. 19.2	Resistance required	Slag aggregate will be free from iron disintegration; the results are declared.
Volume stability of steel slag aggregate, according to EN 1744-1, p. 19.3	V_X	Test applied to basic oxygen furnace slag and electric arc furnace slag; category V_X, where X mean maximum allowed expansion by volume percentage
Water content (added filler), according to EN 1097-5, %	—	Mass content of water in added filler (commercially produced) is fixed and will be maximum 1%.
Stiffening properties: Voids of dry compacted filler (Rigden), according to EN 1097-4	$V_{X/Y}$	The range of Rigden voids in dry compacted filler; categories are labeled as $V_{X/Y}$, where X is a lower limit and Y is an upper limit of voids; note that these are voids according to Rigden's method not Rigden's method modified by Anderson
Stiffening properties: Delta ring and ball, according to EN 13179-1:	$\Delta_{R\&B}X/Y$	The range of increase of softening point (SP) with ring-and-ball method; categories are labeled as $\Delta_{R\&B}X/Y$, where X is a lower limit and Y is an upper limit of SP increase
Water solubility, according to EN 1744-1	WS_X	The water solubility is labeled as WS_X, where X is a maximum allowed percentage (by mass).
Water susceptibility, according to EN 1744-4	—	No specified limits; the results are declared
Calcium carbonate content of limestone filler aggregate, according to EN 196-21	CC_X	Calcium carbonate content is labeled as CC_X, where X is a minimum required percentage (by mass) of $CaCO_3$.
Calcium hydroxide content of mixed filler according to EN 459-2	KaX	Calcium hydroxide (hydrated lime) content is labeled as KaX, where X is a minimum required percentage (by mass) of $Ca(OH)_2$.
Bitumen number of added filler, according to EN 13179-2	$BN_{X/Y}$	The range of bitumen number; categories are labeled as $BN_{X/Y}$, where X is a lower limit and Y is an upper limit

Note: AAV = Aggregate abrasion value; FI = flakiness index; LA = Los Angeles; PSV = polished stone value; SZ = Schlagzertrümmerungswert; SI = shape index.

TABLE 5.2
Requirements for SMA Coarse Aggregate according to EN 13043 in Selected CEN-Member Countries, Aggregates for SMA at the Highest Traffic Level (Reference Mixture SMA 0/11)

Properties[a]	Germany TL Gestein StB 04 Anhang F and TL Asphalt StB 07	Slovakia KLK 1/2009	Austria ONORM B 3584:2006 RVS 08.97.05:2007 (Class G1)	Switzerland SN 670130a: 2005	Poland WT-1 Kruszywa 2008
List of Categories by Country	1	2	3	4	5
Grading, according to EN 933-1	G_c90/10 for (2/5 mm) G_c90/15 for (2/5, 5/8, 8/11 mm)	G_c90/10	G_c90/15	G_c85/15	G_c90/15
Tolerances of typical gradation	—	Declared	—	$G_{20/15}$	$G_{25/15}$
Fines content, according to EN 933-1	Fractions 2/5 to 8/11 mm: f_2 (max 2%)	f_1 (max 1%)	f_1 (max 1%)	f_1 (max 1%)	f_2 (max 2%)
Fines quality, according to EN 933-9—methylene blue test	Declared (value to be reported)	—	—	—	—
Shape of coarse aggregate (Flakiness Index EN 933-3, Shape Index EN 933-4)	SI_{20} or FI_{20} (max 20%)	SI_{20} or FI_{20} (max 20%)	SI_{15} (max 15%)	FI_{25} (max 25%)	SI_{20} or FI_{20} (max 20%)

(Continued)

TABLE 5.2 (CONTINUED)

Requirements for SMA Coarse Aggregate according to EN 13043 in Selected CEN-Member Countries, Aggregates for SMA at the Highest Traffic Level (Reference Mixture SMA 0/11)

Properties[a]	Germany TL Gestein StB 04 Anhang F and TL Asphalt StB 07	Slovakia KLK 1/2009	Austria ONORM B 3584:2006 RVS 08.97.05:2007 (Class G1)	Switzerland SN 670130a: 2005	Poland WT-1 Kruszywa 2008
			List of Categories by Country		
	1	2	3	4	5
Percentage of crushed and broken surfaces in coarse aggregates, according to EN 933-5	$C_{100/0}$, $C_{95/1}$, $C_{90/1}$	$C_{100/0}$	$C_{100/0}$	$C_{95/1}$	$C_{100/0}$
Resistance to fragmentation (crushing), according to EN 1097-2 clause 5 (LA method)	LA_{20} (max 20%)	LA_{25} (max 25%)	LA_{20} (max 20%)	4/8 mm—LA_{25} 8/11 mm—LA_{20} 11/16 mm—LA_{25}	LA_{20} or LA_{25} depending on petrographic type of aggregate
Resistance to fragmentation (crushing), according to EN 1097-2 clause 6 (German *Schlagzertrümmerungswert*)	SZ_{18} (max 18%)	—	—	—	—
Resistance to polishing, according to EN 1097-8	PSV_{51} (min 51)	PSV_{56} (min 56)	PSV_{50} (min 50)	PSV_{50} (min 50)	PSV_{50} (min 50)
Resistance to surface abrasion, according to EN 1097-8 Annex A	—	—	—	—	—

Property					
Resistance to wear (Micro-Deval), according to EN 1097-1	—	$M_{DE}20$ (max 20%)	—	—	—
Resistance to abrasion from studded tires, according to EN 1097-9	—	—	—	—	—
Water absorption according to EN 1097-6	$W_{cm}0.5$ (max 0.5%)	$WA_{24}1$ $W_{cm}0.5$	$WA_{24}1$	Declared	$WA_{24}1$ $W_{cm}0.5$
Resistance to freezing and thawing, according to EN 1367-1 (in water or salt solution)	F_1 (max 1.0%)	F_2 (max 2.0%)	F_1 (max 1.0%)	—	$F_{NaCl}7$ (max 7% in 1% NaCl solution)
Resistance to freezing and thawing, according to EN 1367-2	—	MS_{18} (max 18%)	—	—	—
Resistance to thermal shock, according to EN 1367-5	Declared	—	—	—	—
Affinity of coarse aggregates to bituminous binders, according to EN 12697-11	Declared	—	Min 85% method B	Declared	—
"Sonnenbrand" of basalt, according to EN 1367-3 and EN 1097-2	SB_{SZ} (SB_{LA})	—	SB_{LA}	—	SB_{LA}
Coarse lightweight contaminators according to EN 1744-1 p.14.2	$m_{LPC}0.1$ (max 0.1%)	$m_{LPC}0.1$ (max 0.1%)	—	$m_{LPC}0.1$ (max 0.1%)	$m_{LPC}0.1$ (max 0.1%)

(Continued)

TABLE 5.2 (CONTINUED)
Requirements for SMA Coarse Aggregate according to EN 13043 in Selected CEN-Member Countries, Aggregates for SMA at the Highest Traffic Level (Reference Mixture SMA 0/11)

	Germany TL Gestein StB 04 Anhang F and TL Asphalt StB 07	Slovakia KLK 1/2009	Austria ONORM B 3584:2006 RVS 08.97.05:2007 (Class G1)	Switzerland SN 670130a: 2005	Poland WT-1 Kruszywa 2008
			List of Categories by Country		
Properties[a]	1	2	3	4	5
Dicalcium silicate disintegration of air-cooled blast furnace slag, according to EN 1744-1, p. 19.1	Resistance required	—	Resistance required	According to other regulations	Resistance required
Iron disintegration of air-cooled blast furnace slag, according to EN 1744-1, p. 19.2	Resistance required	—	Resistance required	According to other regulations	Resistance required
Volume stability of steel slag aggregate, according to EN 1744-1, p. 19.3	$V_{3.5}$ (max 3.5%)	$V_{3.5}$ (max 3.5%)	$V_{3.5}$ (max 3.5%)	According to other regulations	$V_{3.5}$ (max 3.5%)

Note: Cells with — mean no requirement (NR) category; FI = flakiness index; LA = Los Angeles; SI = shape index.
[a] Names of properties after EN 13043

TABLE 5.3

Requirements for SMA Fine Aggregate according to EN 13043 in Selected CEN-Member Countries, Aggregates for SMA at the Highest Traffic Level

Properties	List of Categories by Country				
	Germany TL Gestein StB 04 Anhang F and TL Asphalt StB 07	Slovakia KLK 1/2009	Austria ONORM B 3584:2006 RVS 08.97.05:2007 (Class G1)	Switzerland SN 670130a: 2005	Poland WT-1 Kruszywa 2008
	1	2	3	4	5
Grading, according to EN 933-1	G_f85	G_f85	G_f85	G_f85	G_f85
Tolerances of typical gradation	G_{TC}NR	G_{TC}20	G_{TC}20	G_{TC}10	G_{TC}20
Fines content, according to EN 933-1	Declared	f_{10} (max 10%)	f_{16} (max 16%)	f_{22} (max 22%)	f_{16} (max 16%)
Fines quality, according to EN 933-9	Declared	MB_f10 (max 10 g/kg)	—	—	MB_f10 (max 10 g/kg)
Angularity, according to EN 933-6, p.8	E_{cs}35 (min 35 sec)	—	E_{cs}35 (min 35 sec)	Declared	E_{cs}30 (min 30 sec)
Coarse lightweight contaminators, according to EN 1744-1, p.14.2	m_{LPC}0.1 (max 0.1%)	—	—	—	m_{LPC}0.1 (max 0.1%)

TABLE 5.4

Requirements for SMA Filler according to EN 13043 in Selected CEN-Member Countries, Aggregates for SMA at the Highest Traffic Level

Properties	Germany TL Gestein StB 04 Anhang F and TL Asphalt StB 07	Slovakia KLK 1/2009	Austria ONORM B 3584:2006 RVS 08.97.05:2007	Switzerland SN 670130a: 2005	Poland WT-1 Kruszywa 2008
			List of Categories by Country		
Grading, according to EN 933-10			According to Table 24 of standard: Sieve 2.0 mm = 100% passing Sieve 0.125 mm = 85–100% passing Sieve 0.063 mm = 70–100% passing		
Harmful fines (fines quality), according to EN 933-9	Declared			—	$MB_F 10$ (max 10 g/kg)
Water content (added filler), according to EN 1097-5, %	≤%	≤1%	≤1%	≤1%	≤1%
Stiffening properties: voids of dry compacted filler (Rigden), according to EN 1097-4	$V_{28/45}$ (min 28%, max 45%)	—	$V_{28/38}$ (min 28%, max 38%)	$V_{28/45}$ (min 28%, max 45%)	$V_{28/45}$ (min 28%, max 45%)
Stiffening properties: "Delta ring and ball," according to EN 13179-1	$\Delta_{R\&B}8/25$ (min 8°C, max 25°C)	$\Delta_{R\&B}8/16$ (min 8°C, max 16°C)	—	$\Delta_{R\&B}8/25$ (min 8°C, max 25°C)	$\Delta_{R\&B}8/25$ (min 8°C, max 25°C)

Property					
Water solubility, according to EN 1744-1	WS_{10} (max 10%)	WS_{10} (max 10%)	—	Declared	WS_{10} (max 10%)
Water susceptibility, according to EN 1744-4	Declared	—	—	Declared	—
Calcium carbonate content of limestone filler aggregate, according to EN 196-21	CC_{70} (min 70%)	CC_{90} (min 90%)	CC_{80} (min 80%)	Declared	CC_{70} (min 70%)
Calcium hydroxide content of mixed filler, according to EN 459-2	Declared	—	K_a20 (min 20%)	To be established in contract	Declared
Bitumen number of added filler, according to EN 13179-2	—	—	$BN_{28/39}$ (min 28 max 39)	—	Declared

requirement for tolerance of gradation is not necessary; this is the situation in Germany, Switzerland, and a few other countries. In Poland, where limits for gradation during aggregate production do not exist, it was necessary to put such a requirement (categories G and G_{TC}) in the national specifications.

• Legal systems and approaches to the requirements' system also play roles. In Europe, aggregates for asphalt mixtures are construction products and are produced and placed on the market according to Construction Product Directive* regulations. Internal regulations of each country have to be consistent with this directive. The product (aggregates) must fulfill specified requirements for intended use; countries are free to determine how they specify the system of requirements (only a few properties are indispensable). When we see Tables 5.2 through 5.4, it is obvious that some countries built a very broad system and established more detailed specifications than others. The reason is most likely in the existing approach to the requirements for components. In some countries only a few properties are specified because the final mixture (e.g., SMA) features are treated as the most important and a large degree of freedom is left for asphalt mix producers as long as they meet the final desired properties. In other countries everything is specified—both components and final mixture properties as well, which can ultimately lead to overspecification.

5.1.2 UNITED STATES

U.S. requirements for aggregates constitute a compromise between high quality conditions and the necessity of taking into account the economics of manufacturing asphalt mixtures. The number of properties specified are limited, while the requirements themselves are somewhat broad (see Table 5.5), compared with the European standards. The requirements also vary from state to state. Similar requirements are used in the United States for SMA airfield surfacing (ETL 04-8).

Additionally, the possibility of using reclaimed dusts (baghouse fines) from an asphalt mixing plant is a fine example of the pragmatic approach to the selection of aggregates for asphalt mixtures.

Dusts, mineral powders, hydrated lime, and pulverized fly ashes are allowed, while lumps and organic impurities are excluded as fillers in the United States.

5.2 REQUIREMENTS FOR BINDERS

SMA mixtures are chiefly laid as wearing courses. Binders for them should therefore have suitable properties for asphalt mixtures applied to that layer.

The majority of SMAs are placed in moderate climates. Therefore the SMA binder is usually an unmodified binder, or sometimes a polymer-modified one, with a penetration between 50 and 100 (0.1 mm) at 25°C. In several countries, multigrade

---

* Council Directive 89/106/EEC of 21 December 1988 on the approximation of laws, regulations, and administrative provisions of the Member States relating to construction products.

TABLE 5.5

Requirements for SMA Aggregates (Coarse, Fine, and Filler) in the United States

Properties	Requirement	Comments
	Coarse Aggregates	
Crushing resistance: LA abrasion test, % loss (AASHTO T 96)	≤30	There is a suggestion to use additional test methods like Micro-Deval or SGC degradation test. Despite experiences with aggregates of LA 30–45%, their use is not recommended due to possible grain crushing during compaction both in the laboratory and on the road.
Shape of particles: Flat and elongated, % (ASTM D 4791)	≤20 at 3:1, ≤5 at 5:1	Aggregates that have a high percentage of flat and elongated particles: • Tend to break down during compaction • Have higher voids within the aggregate Requirements apply to a whole coarse aggregates fraction in SMA but not to individual materials (fractions).
Particle surface. Crushed content:, % (ASTM D 5821)	100% one face, ≥90% two faces	Property is important for interlocking of aggregates skeleton.
Absorption in water, % (AASHTO T 85)	≤2	—
Susceptibility to weathering: Soundness, 5 cycles, % (AASHTO T 104)	≤15 in sodium sulfate, ≤20 in magnesium sulfate	Tests show that there is good correlation between magnesium sulfate soundness and micro-Deval abrasion test.
	Fine Aggregates	
Susceptibility to weathering: Soundness, 5 cycles, % (AASHTO T 104)	≤15 in sodium sulfate, ≤20 in magnesium sulfate	See Coarse Aggregates.
Angularity, % (AASHTO T 304, method A)	≥45	Indication of interlocking potential of fine aggregates
Atterberg Limits Tests		
Liquid limit, % (AASHTO T 89)	≤25	The liquid limit is the water content at which the material passes from a plastic to a liquid state. The plasticity index is the numerical difference between the liquid limit and the plastic limit; it is the moisture content at which the material is in a plastic state. The goal is to eliminate aggregates with clay or silt particles.
Plasticity index (AASHTO T 90)	Nonplastic	

(Continued)

TABLE 5.5 (CONTINUED)
Requirements for SMA Aggregates (Coarse, Fine, and Filler) in the United States

Properties	Requirement	Comments
Absorption, % (ASTM C 128)	≤2	Used in airfields [ETL 04-8]
Sand equivalent, % (ASTM D 2419)	≥45	Used in airfields [ETL 04-8]; the goal is to eliminate aggregates with clay or silt particles. Shows the relative proportion of plastic fines (and dust) to sand fraction.
Filler		
Plasticity index (AASHTO T 90)	<4	See Fine Aggregates.
Modified Rigden voids content, %	<50	Recommended value

Note: AASHTO = American Association of State Highway and Transportation Officials; ASTM = American Society for Testing Materials; LA = Los Angeles; SGC = Superpave Gyratory Compactor; SMA = stone matrix asphalt.

binders are also used. Under special circumstances, especially in countries with a cold climate, soft binders (with penetration higher than 100 at 25°C) are used. A steady increase in the percentage of polymer-modified binders in SMA mixtures has been observed recently. Considerable research (see Chapter 12) has proved that a polymer-modified binder substantially improves the characteristics of a finished SMA layer.

5.2.1 REQUIREMENTS ACCORDING TO EUROPEAN STANDARDS

In Europe, road binders (paving grade) according to EN 12591 and modified binders according to EN 14023 have been used in SMA mixtures. Normally, road binders of 50/70 and 70/100 are used; however softer 160/220 types have also been cited in the regulations for roads with light traffic. The properties of each type of paving grade binder are fully specified in EN 12591; Table 5.6 shows selected properties of 50/70, 70/100, and 160/220.

European Standard EN 14023 for polymer modified binder (PMB) is similar to EN 13043 for aggregates. It comprises two tables with PMB properties to be chosen by any CEN country. In this way, PMB with the same designation could vary significantly from one country to another. Table 5.7 provides examples of some PMBs specified for SMA.

Recently, a technology utilizing the simultaneous application of modified binder with a viscosity-reducing additive at high temperatures (e.g., F–T wax) has achieved great popularity.

TABLE 5.6

Selected Properties of Paving Grade Binders (Unmodified) according to EN 12591

		Paving Binder Grade		
Properties	Method	50/70	70/100	160/220
Penetration at 25°C, 0.1 mm	EN 1426	50–70	70–100	160–220
Softening point R&B, °C	EN 1427	46–54	43–51	35–43
Fraass breaking point, °C	EN 12593	≤ -8	≤ -10	≤ -15
Retained penetration at 25°C after RTFOT, %	EN 12607-1, EN 1426	≥ 50	≥ 46	≥ 37

Note: R&B = Ring and Ball; RTFOT = Rolling Thin Film Oven Test.

5.2.2 UNITED STATES

The requirements for SMA binders used in the United States are related to the Superpave system prevailing in the country. The basis for the selection of binders for wearing courses, in principle their functional type (performance grade [PG]), is the climate of the given area of application. So all binders have to meet the same requirements for deformation resistance, fatigue, and low temperature cracking, but they meet those limits at different temperatures that relate to the climate in various parts of the United States. In specified situations, such as low-speed traffic or very high-traffic loads, the system additionally allows increasing the upper (high temperature) grade of the binder. Finally, a binder that meets the specific climatic and traffic requirements for the construction location is then specified or selected.

So there are no special nationwide requirements for SMA binders. The rules for binder selection from Superpave are generally followed for all mixtures. Readers interested in the PG system can find detailed information in SHRP or Asphalt Institute publications (e.g., *SP-1 Superpave Performance Graded Asphalt Binder Specifications and Testing*).

5.3 REQUIREMENTS FOR STABILIZERS

Specifying the desired properties of stabilizers is a troublesome task, which is why there are not very many examples of formalized requirements for them. In fact, detailed testing of a stabilizer's properties is practically impossible in the asphalt plant laboratory. Therefore, in many countries, testing is limited to draindown checking (i.e., only an empirical assessment of stabilizer effectiveness).

A few examples of standardized regulations come from Germany and the United States. Some requirements were also in the Finnish specification PANK 2000.

5.3.1 GERMANY

The requirements adopted in Germany were cited in a 1997 document entitled *Testing and Marking Stabilizing Additives and Materials Applied for Bituminous*

TABLE 5.7

Examples of Requirements for Selected PMB according to EN 14023 in Selected CEN-Member Countries

Properties	Test Method	GERMANY TL-Bitumen 07 25/55-55	SWEDEN VV Publ 2008:113 40/100-75	POLAND PN-EN 14023:2009/ AC2010 45/80-55	SLOVAKIA Katalógové Listy Asfaltov KLA 1/2009 45/80-55
Penetration at 25°C, 0.1 mm	EN 1426	25–55	40–100	45–80	45–80
Softening point, °C	EN 1427	≥55	≥75	≥55	≥55
Force ductility (low speed traction), J/cm²	EN 13589 EN 13703	≥3 at 5°C	TBR	≥3 at 5°C	≥2 at 5°C
Change of mass, %	EN 12607-1	≤0.5	≤0.5	≤0.5	≤0.5
Retained penetration after RTFOT, %	EN 1426	≥60	≥60	≥60	≥60
Increase in softening point after RTFOT, °C	EN 1427	≤8	≤10	≤8	≤12
Flash point, °C	EN ISO 2592	≥235	≥220	≥235	≥250
Fraass breaking point, °C	EN 12593	≤−10	≤−12	≤−12	≤−18
Elastic recovery at 25°C, %	EN 13398	≥50	—	≥50	≥50
Elastic recovery at 10°C, %	EN 13398	—	≥75	—	—
Plasticity range, °C	p. 5.1.9	—	TBR	TBR	—
Storage stability	EN 13399	—	—	—	—
Difference in softening point, °C	EN 1427	≤5	≤5	≤5	≤5

Property	Test method			
Storage stability	EN 13399	—	—	—
Difference in penetration, 0.1 mm	EN 1426	—	TBR	—
Drop in softening point after RTFOT, °C	EN 12607-1, EN 1427	≤2	≤5	TBR
Elastic recovery at 25°C after RTFOT, %	EN 12607-1, EN 13398	≥50	—	≥50
Elastic recovery at 10°C after RTFOT, %	EN 13398	—	≥50	—

Note: TBR = To Be Reported.

Surfaces (FGSV Arbeitpapier 42). It contains requirements for all kinds of stabilizers used in aggregate mixtures with a high content of binder, which includes the
following:

- Organic fibers
- Mineral fibers
- Powders and dusts
- Compounds of filler and all sorts of fibers
- Special fillers.

The range of testing depends on the type of stabilizer. The following is a set of
recommended tests for fibers:

- Macroscopic assessment
  - Homogeneity
  - Color
  - Odor
  - Agglomeration (balling)
- Microscopic assessment
  - Structure
- Other properties
  - Gradation distribution
  - Diameter—in specified cases
  - Mass loss after drying
  - Mass loss after ignition
  - Specific gravity
  - Water susceptibility
  - Stiffening properties (an increase of the softening point after the
    Wilhelmi method)

5.3.2 UNITED STATES

The United States is one of the few countries where requirements for cellulose fibers
and mineral stabilizing agents have been standardized. Some other countries have
adopted the U.S. requirements in their own specifications.

The requirements for cellulose and mineral fiber SMA stabilizers after AASHTO
MP-8-05 include the following:

- Fiber length: maximum 6 mm
- Thickness or diameter of mineral fibers: maximum 0.005-mm mean test
  value
- Gradation of cellulose fibers
  - Passing a 0.15-mm Alpine sieve (method A): 70% ± 10%,
  - Passing 0.85-, 0.425-, and 0.106-mm mesh screen sieves (method B):
    85% ± 10%, 65% ± 10% and 30% ± 10%, respectively

- Shot (nonfibrous material) content
  - Passing 0.25-mm sieve: 90% ± 5%
  - Passing 0.063-mm sieve: 70% ± 10%
- Ash content of cellulose fibers: 18% ± 5% nonvolatiles
- pH of cellulose fibers: 7.5 ± 1.0
- Oil absorption of cellulose fibers: 5.0 ± 1.0 times fiber mass
- Moisture content of cellulose fibers: maximum 5% (m/m)

5.3.3 FINLAND

The requirements for cellulose fibers according to PANK 2000 have been divided into obligatory and recommended ones as follows:

- Obligatory requirements
  - Water content according to PANK 3103: ≤ 8.0% (m/m),
  - Instantaneous heat resistance (mass loss) according to PANK 3104: ≤ 7.0% (m/m)
- Recommended values for bulk fibers
  - Bulk density according to PANK 3105: 20–35 g/dm^3
  - Homogeneity according to PANK 3107: 2.0–2.8%
  - Fibers' length distribution:
    – 80% value: 1.2–1.9 mm
    – 50% value: 0.5–0.9 mm
  - Specific surface area according to PANK 2401: 2.0–3.0 m^2/g

5.4 RECLAIMED ASPHALT

Generally, most regulations do not recommend using reclaimed asphalt or simply do not allow recycling old asphalt layers into SMA. However, some good test results do exist for SMA mixtures with reclaimed asphalt (Perez et al., 2004).

The requirements for reclaimed asphalt to be used for SMA according to the EN 13108-5 standard are discussed in Chapter 14.

5.5 OTHER MATERIALS

5.5.1 NATURAL ASPHALT

Some results of the use of natural asphalt as an additive to road binder manufactured in a refinery are cited in relevant literature (Häusler and Arand, undated; Radenberg, 1997). Natural asphalt is usually added to increase resistance to permanent deformation.

The requirements for natural asphalts have also been cited in the EN 13108-4 standard, Annex B.

5.5.2 ARTIFICIAL SLAG AGGREGATES

Slags of different sorts may be used in SMA mixtures, providing they meet suitable requirements. For example, in Europe these can be determined in the National

Application Documents for the EN 13043 standard; examples were presented in Table 5.2 (for more information, see Chapter 14). When considering the possibility of using slag in an asphalt mixture, one should remember the following specific properties for these aggregates:

- The density could sometimes be substantially different from that of crushed rock aggregates and can be variable; when designing an SMA mixture using the weight method, differences can manifest themselves in differing volume relations between mixture components.
- The binder content should be determined carefully; a higher density of an aggregate mix usually brings about the necessity for reducing the quantity of binder, but the high porosity of external surfaces of slag grains should also be taken into account.
- The complex chemical composition of slag means that adhesion promoters should be carefully selected and confirmed with testing.
- One of the basic requirements imposed on slag aggregates is an invariability of properties determined during testing called chemical disintegration.

In addition to these properties, some research (Airey et al., 2004) has proven that mixtures with slag aggregates demonstrate very good interlocking characteristics and a relatively high resilient modulus. Conversely, mixtures with slag showed an increased susceptibility to age hardening (in long-term laboratory aging).

Other tests of slag aggregate show very high PSV values, which could be used for designing skid-resistant asphalt mixes.

6 Designing SMA Composition

Stone matrix asphalt (SMA) mixtures and courses made from them have many strengths. Naturally, these mixtures must be well-designed, but that attribute has various shades of meaning. A review of publications on this subject has not revealed a method that is clearly "the best." There is as wide a variety of design methods as there are approaches to the roles of particular constituent materials.

Having decided to place such a mixture, a civil engineer faces a challenge. It is not an easy material to deal with, neither during design nor construction. The first essential task is to achieve the proper SMA composition. All the remaining aspects of SMA construction—namely, production, transportation, placement, and compaction—are affected by this first step. Most problems at subsequent stages of work with SMA can be avoided only by achieving a good mix design.

In the various sections of this chapter, the following issues will be presented and discussed:

- Selection criteria for an aggregate mix size (the maximum particle size) depending on the design thickness of the course
- Design method with gradation limits
  - Designing the part of the aggregate mix greater than 2 mm
  - Designing the part of the aggregate mix less than 2 mm
- Selection of the proper binder content
- Final assessment of the properties of the designed mixture

Chapter 7 presents an overview of selected SMA design methods developed in various countries.

6.1 SELECTING SMA GRADATION AND SIZE

No matter the method applied when designing the SMA composition, the first step that must be taken is establishing the maximum particle size in the aggregate mix. The following factors should be taken into account when considering this issue:

- SMA position (in a wearing course or in an intermediate course)
- Design thickness of the course after compaction
- Traffic load and the location of the road section (e.g., rural or urban)
- Additional requirements for the SMA course

6.1.1 IN WHICH COURSE?

Due to the outstanding performance of SMAs in surface courses, they have also been used in intermediate courses (e.g., in the United States, Australia, and recently Germany). Normally mixtures with larger maximum particle sizes (e.g., 0/19 or 0/22 mm) are selected for that purpose when they are permitted by local regulations. So far, aside from possible economic obstacles (high cost), it has not been stated that SMA is unfit to be used in intermediate courses. Consequently, if one can afford to and knows how to design and execute SMA intermediate courses, why not dare to?

When using SMA as a course on a bridge deck, the gradation should be selected rather conservatively (i.e., it should be finer). The composition of the asphalt mixture should also be designed with great care and mostly with the use of the type of polymer modified binders that increase the fatigue life of an asphalt mix. (See Chapter 13 for further discussion of SMA for bridge decks.)

6.1.2 WHAT THICKNESS OF A COURSE?

In bituminous mixtures, the adopted rule used to be that the thickness of a course should not be less than 3.5 times the nominal maximum aggregate size (NMAS) in a given mixture. However, due to problems with compaction, usually 3.5–4.0 times the NMAS in a mixture is normally suggested as the appropriate thickness of an SMA course. A course that is too thin in comparison with the maximum particle size causes the following:

- Tearing the mat during laydown and cracking during rolling
- Problems with compacting the course
- Breaking of weaker particles during rolling
- Problems with the appropriate arrangement of particles and weakening of the aggregate structure

Many SMA guidelines provide for a range of thicknesses for each particular mixture. Some regulations from the old German guidelines ZTV Asphalt-StB 01 and from the new ZTV Asphalt-StB 07 on the thickness of the SMA 0/11 course are noteworthy because of an exceptionally narrow range of placement thickness for this mixture, specifically 35–40 mm (Table 6.1).

In some countries, a mass criterion on 1 m^2 of the surface area (spread rate) is specified instead of designating the recommended thickness of a course. Knowing the SMA bulk density, the thickness of a layer of that mixture may be calculated. The German ZTV Asphalt-StB 01 and ZTV Asphalt-StB 07 and Finnish regulations PANK 2008 are examples of such approaches (Table 6.2). Table 6.1 shows particle sizes and corresponding courses according to various guidelines.

Finally, it is worth noting that a thin course gets cold much faster than a thick one, which has a significant effect on compaction during the cooler periods of the year.

TABLE 6.1
Thicknesses of SMA Wearing Courses, depending on the Maximum Particle Size according to Various Guidelines for SMA

					Thickness of a Compacted Course in Different Countries, mm			
Mixture	Austria RVS 08.16.01	Slovakia KLAZ 1/2008	Poland WT-2- 2008	United States NAPA IS 128[a]	Germany ZTV Asphalt- StB 07	Germany ZTV Asphalt- StB 01	United Kingdom BS 594987:2007
SMA 0/5	—	—	20–40	—	20–30 (N)	20–40	20–40 (SMA 0/6)
SMA 0/8	25–35	20–40	25–50	25–37.5 (SMA 0/9.5)	30–40 (S) 20–35 (N)	30–40 (S)[b] 20–40	—
SMA 0/11	30–40	30–50	35–50	37.5–50 (SMA 0/12.5)	35–40 (11S)	35–40 (11S)[b]	25–50 (SMA 0/10)
SMA 0/16	—	40–60	—	50–75 (SMA 0/19)	—	—	35–50 (SMA 0/14)

Note: N = low/medium traffic; S = heavy traffic.

[a] In NAPA IS 128 publication, only recommended minimum lift thicknesses ranges are presented.

[b] In special circumstances, when the SMA bottom layer is not even, the following course thicknesses have been allowed: of SMA 0/11S from 25 to 50 mm, and of SMA 0/8S from 20 up to 40 mm.

TABLE 6.2
The Mass of SMA Wearing Courses, depending on the Maximum Particle Size[a]

Guidelines	Mass of 1 m² of a Surface Area, kg					
	SMA 0/22	SMA 0/16	SMA 0/11S	SMA 0/8S	SMA 0/8	SMA 0/5
German guidelines ZTV Asphalt-StB 01	—	—	85–100	70–100	45–100	45–75
German guidelines ZTV Asphalt-StB 07	—	—	85–100	75–100	50–85	50–75
Finnish guidelines PANK 2008	100–150	80–125	60–100	—	50–90	50–75

[a] According to the German ZTV Asphalt-StB 01 and ZTV Asphalt-StB 07 and Finnish regulations PANK 2008.

6.1.3 TRAFFIC LOADING AND LOCATION

Coarse-graded mixtures make stronger skeletons. That is why the majority of requirements for SMA contain a noticeable tendency toward increasing the maximum particle size in a mixture in conjunction with an increase in the traffic loading. When selecting a mixture, both the strengths and weaknesses of an accepted solution should be considered; mixtures with larger maximum particle size (let us suppose greater than 10 mm) are characterized by higher rut resistance but lower noise reduction ability and poorer skid resistance. Therefore, when at all possible, SMA 0/11 is being gradually abandoned for SMA 0/8.

In Germany, SMA 0/8S and 0/11S have been used on the most heavily trafficked roads while on lesser trafficked roads mixtures without an S marking are used. The guidelines from 2001 to 2007 (ZTV Asphalt-StB 01 and ZTV Asphalt-StB 07, respectively) have differentiated mixtures according to traffic, with the understanding that mixtures with the same maximum aggregate size but intended for various traffic loadings differ in the shape of the gradation limits. German SMA aggregate mixes of the S type differ from the lower traffic SMAs in that they have the following:

- Lower filler contents
- Lowered gradation limit curves
- No non manufactured (natural) sand

As a result, German SMA mixtures of the "S" type should be coarser and, at the same time, less closed. This is logical since there should be more voids in a heavily trafficked course (refer also to the Dutch method in Chapter 7).

In many countries, as in Germany, different SMA gradation curves have been specified depending on the traffic category (traffic loading), as shown in Table 6.3.

TABLE 6.3
SMA Division according to the Traffic Category Prevailing in Locations

	SMA Mixes with Gradation						
Application	Germany ZTV Asphalt-StB 07	United States NAPA IS 128	Canada (Ontario) OPSS. MUNI 1151	Slovakia KLAZ 1/2008	Poland WT-2 2008	Slovenia SIST 1038-5:2008	Australia NAS 2Ed 2004
Low and medium traffic roads	0/8N, 0/5N	—	—	—	0/5, 0/8	0/4, 0/8, 0/11	0/7, 0/10
Principal roads and highways (heavy traffic)	0/11S, 0/8S	0/9.5, 0/12.5, 0/19	0/9.5, 0/12.5, 0/19	0/8, 0/11, 0/16	0/8, 0/11	0/8, 0/11	0/10, 0/14

6.1.4 ACCORDING TO THE EUROPEAN STANDARD EN 13108-5

The requirements for gradation of SMA mixtures have been provided in the European standard PN-EN 13108-5 (see Chapter 14). This standard does not set out the criteria and conditions for selecting the particular gradation of a mixture. Establishing these criteria remains the responsibility of each CEN-member state. In Figures 14.2 through 14.5, examples of gradation limits are presented.

6.2 GENERAL RULES

6.2.1 ORIGINAL ZICHNER'S PROPORTIONS

In his publications (Zichner, 1971; Zichner, 1972) Dr. Zichner described a recommended composition of an SMA mixture as follows:

- The stone content should be about 65–80% (m/m), preferably 70–75%, using only crushed stones.
- The main rule governing gradations is that the mixture is composed "so that the percentage of the coarser size is greater than that of the smaller size."
- Mastic content is 20–35% (m/m).
- 23–28% of the mastic is a binder.
- For layers with different thicknesses, different types of stones should be used as follows:
  - Thickness less than 3 cm—stone fractions 2/5.6 and 5.6/8 mm in proportions of 25 and 75%, respectively
  - Thickness 3–4.5 cm—stone fractions 2/5.6, 5.6/8, and 8/12 mm, or only 5.6/8 and 8/12 mm
  - Thickness greater than 4.5 cm—stone fractions 2/5.6, 5.6/8, 8/12, and 12/18 mm
- Only manufactured sand should be used in the mastic.

Looking at Zichner's MASTIMAC and MASTIPHALT gradation curves presented in Chapter 1 (see Figure 1.1), we can find approximate contents of aggregate fractions (Table 6.4).

The proportions of different fractions of coarse aggregates used in current German SMA mixtures are presented also in Chapter 2 (see Table 2.1) after the German DAV Handbook (Drüschner and Schäfer, 2000).

6.2.2 30–20–10 RULE

The 30–20–10 rule suggests that proper stone-to-stone contact is created if the percentages of aggregate passing sieves of 0.075 mm, 2.36 mm, and 4.75 mm equal 10%, 20%, and 30%, respectively, which should provide for the appropriate discontinuity in the gradation. After comparing this rule with the data of Table 6.4, one can see that the proportion of grains larger than 4.75 mm from this rule (70%, or 30% passing the

TABLE 6.4
Approximate Contents of Aggregate Fractions for Zichner's Mastimac and Mastiphalt

Mixture	Filler Fraction < 0.09 mm	Sand Fraction 0.09–2.0 mm	Aggregate 2/5.6 mm	Aggregate 5.6/8 mm	Aggregate 8/12.5 mm
MASTIMAC (SMA 0/8)	12–13%	11–12%	15%	60%	—
MASTIPHALT (SMA 0/12.5)	12–13%	11–12%	10%	27%	38%

Source: Based on Zichner, G., MASTIMAC unad MASTIPHALT bituminöse Gemische für hochwertige Deckschichten. STRABAG Schriftenreihe 8, Folge 4, 1972.

4.75-mm sieve) is a bit more than Zichner assumed for his MASTIPHALT (65%). For the other fractions (i.e., those for filler and sand) the rule gives the opposite effect—a reduction in the quantities of those materials. In summary, the 30–20–10 rule produces a coarser SMA mixture and with lesser amounts of sand and filler.

Some of the first SMA sections in the United States were designed according to this rule (Scherocman, 1991).

6.2.3 BINARY SYSTEMS

Changes in the content of air voids in an aggregate mix are consequences of changes in the aggregate gradation. The relationships between the content of the coarse aggregate and the content of air voids have been known for many years; a discussion of them may be found in the publication by Schroeder and Kluge (1992). The same applies to the impact of the coarse aggregate to fine aggregate ratio on air voids (so-called binary systems). A relationship of this type is shown in Figure 6.1 (Ferguson et al., 1999; Francken and Vanelstraete, 1993; Lees, 1969; Voskuilen, 2000).

When considering the test results of example binary systems, the following can be found:

- If the fine aggregate constitutes 100% of a mixture, a certain level of air voids among fine grains will be its distinctive feature.
- When adding coarse grains to the fine aggregate, the content of voids among the grains becomes gradually smaller, which can be explained as replacing some volume of sand grains and the voids among them with a volume of solid coarse grains; that process may be called a *replacement* or *substitution phase* (Francken and Vanelstraete, 1993; Voskuilen, 2000) (see Chapter 7).
- With the content of the coarse aggregate at 60–70% (m/m), the content of voids in an aggregate mix reaches its minimum; it is impossible to pack in more coarse grains since all of them have already come into contact with

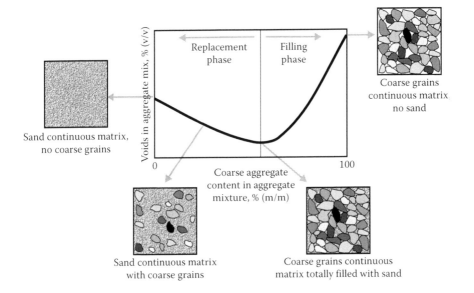

FIGURE 6.1   Relationship between contents of voids in the aggregate mix and the coarse aggregate fraction. (Based on Ferguson, A., Fordyce, D., and Khweir, K., *Proceeding of the Third European Symposium on Performance and durability of bituminous Material and Hydraulic Stabilised Composites*, AEDIFCATIO publishers, D-79104 Freiburg i. Br. and CH-8103 Unterengstringen/Zurich, 1999; Francken, L. and Vanelstraete, A., *Proceeding of Eurobitume Congress Stockholm*, Sweden, 1993; Lees, G., *Journal of the Association of Asphalt Paving Technologists*, 39, 1969; van de Ven, M.F.C., Voskuilen, J.L.M., and Tolman, F., The Spatial approach of hot mix asphalt. *Proceedings of the 6th RILEM Symposium PTEBM'03*, Zurich 2003.

each other (a skeleton of coarse grains has been formed and is filled with fine grains); the *replacement phase* has come to an end.

- Then follows the reverse direction of changes in the contents of voids—the mixture becomes open by means of gradually removing the fine aggregate among coarse grains up to 100% (m/m) coarse aggregate when the highest content of voids is reached; that process could be named the *filling phase*.

The aforementioned relationship between the amount of air voids and gradation of the coarse aggregate fraction directly translates into the binder content in SMA (Drüschner and Harders, 2000; Schroeder and Kluge, 1992). The difference in the binder content, which is dependent on the content of grains larger than 2 mm, has been proved in the previously mentioned German publications (and many others published in Germany). In these examples of tested SMA mixes, the optimum binder content in an SMA mixture depended on the coarse fraction content. For example, with 73% (m/m) of coarse particles content and air voids at the level of 3% (v/v), the binder content amounts to 5.5% (m/m); after an adjustment of the aggregate mix and an increase in the content of coarse aggregate up to 80% (m/m), the same 3% (v/v) of air voids are achieved at a binder content of about 7% (m/m). These results were achieved with the use of a Marshall hammer with a compaction effort of 2 × 50 strokes.

This relationship among contents of binder, air voids, and the coarse aggregate fraction establishes a rule that the content of voids in a designed SMA mixture should not be adjusted by changing the binder content. There is a much higher potential for changing the air voids by adjusting the content and gradation of the coarse aggregate fraction or, generally, by altering the gradation curve.

6.3 DESIGNING AN AGGREGATE MIX BY APPLYING GRADATION LIMITS

The application of gradation limits has become the most commonly used method for designing SMA mix composition. This method involves gradation analyses of all the constituent aggregates, including the filler, followed by balancing the proportions of all the aggregates in such a way that the ultimate gradation curve is situated between the adopted gradation limits. Used alone, it is a very simple method. Unfortunately it is characterized by some drawbacks that may result in a poor SMA mix design.

The advantages of the method of gradation limits include the following:

- The method is simple to use and quick to produce results. After the gradation analyses of the constituent materials are carried out, the proportions of the mix constituents can be quickly calculated simply by using computer software. It is also easy to relocate the gradation curve by manipulating the percentages of the constituents. However, for the very experienced engineer there are better ways to analyze and predict the behavior of the mix.

The disadvantages of this method are as follows:

- The position of a gradation curve inside gradation limits does not absolutely secure an appropriate design of an aggregate mix. For example, the gradation curve graph itself is not enough to predict and secure both a suitable aggregate skeleton and the impact of flat and elongated particles.
- Some major or minor errors occur in volume relations of constituent materials when using only mass units to show the gradation. If the method does not stipulate this, differences in densities of constituents are not taken into consideration. Significant differences in aggregate densities produce substantial differences in the volume relations of a mix. This is not apparent on a gradation curve if it shows the weight distribution retained on sieves, though it can obviously be solved by presenting the gradation in terms of volume units.
- It is presupposed that gradation limits illustrate the area of an appropriate gradation. However, as practice proves, the limits of a suitable gradation should also be periodically verified and mistakes corrected.

In summary, it is safe to say that relying only on the method of gradation limits leads to designing SMA with a low degree of reliability of performance. In Chapter 7, some SMA design methods are discussed, with the application of some concepts extending beyond the use of gradation limits.

6.3.1 DESIGNING A GRADATION CURVE

Some hold the opinion that the best gradation curve is the one passing exactly in the middle of the space between the upper and lower gradation limits. To a certain extent in some cases this may be true; however, in the majority of cases it is not. The shape of a design gradation curve exerts a significant impact on mix properties. For example, by looking at its shape, one may determine if the mix is more or less coarse or has the probability of being overfilled with mastic. Therefore the shape of a design gradation curve is not an unimportant question.

The subsequent phases of design will be discussed later, starting with the coarse aggregate fraction, going through the sand fraction, and ending with the filler content. A familiarity with the basic rules of designing aggregate mixes, including the algorithm and calculations, is assumed. To brush up on these rules, refer to a basic text, such as *The Asphalt Handbook* (MS-4), a publication of Asphalt Institute.

6.3.2 DESIGNING THE AGGREGATE MIX LARGER THAN 2 MM

Let us design the SMA coarse aggregate fraction (particles larger than 2.0 mm). The majority of guidelines recommend its content should be about 70–80% (m/m) of the whole aggregate mix, making it the major component. It may seem that there is little room to maneuver; however, there is a huge potential for controlling the SMA properties by making changes within this narrow range.

Some issues are worth deliberating, namely the impact on a mix exerted by the following:

- A change in the content of coarse aggregates (grains retained on a 2 mm sieve)
- The actual gradation of the coarse aggregate fraction (distribution of coarse aggregate on sieves larger than 2 mm)
- The density of the coarse aggregate particles

To illustrate these issues, two model mixes of coarse aggregates, calculated in a laboratory, will be discussed. One point should be mentioned at the beginning. We will use the following sieve set for coarse fraction: 2.0 mm, 4.0 mm, 5.6 mm, 8.0 mm, 11.2 mm, and 16.0 mm.

6.3.2.1 Part I: Coarse Aggregate Content

Designing the coarse aggregate part of an SMA mix differs widely from designing an asphalt concrete (AC) mix. The unexpected difference lies in the various results of similar actions. Let us consider the example of SMA gradation curves.

It should be kept in mind that SMA has a strong aggregate skeleton with little to none of the medium aggregate fraction. Different distributions of sizes within the coarse aggregate fraction can lead to greater or smaller discontinuities in the overall gradation in a course, which can lead to some pretty interesting consequences for the mixture.

Example I

MIXTURE S

Thus let us take the gradation limits of SMA 0/12.8 mm and insert a gradation curve between them, marking that design as S (Figure 6.2). Grading parameters of the mixture S are shown in Table 6.5.

MIXTURE S1

Let us perform the operation of adding coarse aggregates to the mixture. Now, more coarse graded grains have been added as can best be seen on 2.0- and 4.0-mm sieves. This creates mixture S1 (Figure 6.2). The gradation characteristics of a new mixture S1 may also be found in Table 6.5. The gradation curve for mixture S1 has been moved downward in the area of sieves larger than 2 mm relative to mixture S.

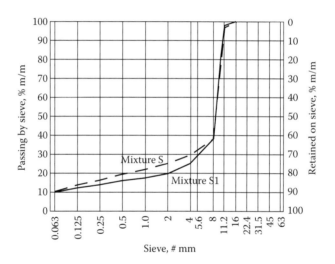

FIGURE 6.2   Grading curves of S (the dotted line) and S1 (the solid line) mixes—SMA of Example I.

TABLE 6.5
Grading Characteristics of Mixtures S and S1

Gradation Properties	Mixture S	Mixture S1
Filler fraction content (grains <0.063 mm), % (m/m)	10.5	10.1
Sand fraction content (grains 0.063–2.0 mm), % (m/m)	14.4	9.8
Coarse aggregate content (grains >2.0 mm), % (m/m)	75.5	80.1
Coarse aggregate content (grains >4.0 mm), % (m/m)	70.0	74.3
Specific surface area of the mixture, cm²/g	189	170

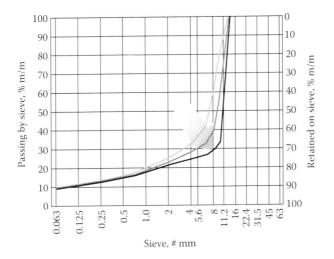

FIGURE 6.3   SMA mixtures with increasingly higher contents of the coarse aggregate fraction.

CONCLUSIONS OF THE EXAMPLE I SMA MIXTURE

Conclusions drawn from the analysis of Example I include the following:

- The increase in coarse grains in a mix brings about the lowering of curves in the area of sieves larger than 2 mm. There then follows an increase in particles retained on the 2-mm sieve; all in all, SMA gradation curves behave the same way as those of AC.
- The main difference in gradation between mixtures S and S1 is that the sand fraction (0.063/2.0 mm) of mixture S1 is less by about 5% than that of mixture S, and this percentage of material is moved to fraction 2.0/8 mm of S1 mixture.
- It is easy to observe that the amount of coarse aggregates of fraction 8/11.2 mm is almost the same in mixtures S and S1; this means that the main changes occurred between sieves 0.063 and 8 mm.
- There are more coarse aggregates in mixture S1, hence the quantity of coarse aggregates carries with it an increase in the air voids in the aggregate mix and an increase in the quantity of binder needed in the SMA. Meanwhile, let us look at Figure 6.3, which shows numerous gradings of SMA mixtures. Each of them contains more and more coarse particles, and the gradation curves move in the direction of the arrow. Naturally, the consequences of such movements are more and more spacious voids among coarse particles.

6.3.2.2   Part II: Gradation within the Coarse Aggregate Fraction

When designing the shape of a gradation curve in the coarse aggregate fraction (larger than 2 mm), it is worthwhile to pay attention not only to the percentage of grains larger than 2 mm (retained on a 2 mm sieve) but also to the ratios of contents of various coarse aggregate fractions. According to many recommendations (i.e.,

Schroeder and Kluge, 1992; Voskuilen, 2000), mixtures characterized by a certain deficiency of smaller coarse aggregates, or even a lack of them, should be preferred. So, let us proceed to Example II to discuss changes in the gradation within the coarse aggregates' fraction. This example is not complicated, therefore we will not encounter any difficulties in its interpretation; this is especially for anyone who often designs using the gradation limits method.

Example II: Changes in Gradation within the Coarse Aggregate Fraction

Let us start analyzing an SMA mixture that has fixed contents of particles larger than 2 mm at 75% (m/m), filler at 10%, and fine aggregate at 15%. The gradation curves of SMA 0/11 according to ZW-SMA-2001, fine aggregate 0/2, a filler, and 3 fractions of the coarse aggregate (2/5.6, 5.6/8, and 8/11.2) will be used for the analysis. To make the task easier, an assumption has been made that none of the aggregate fractions has any undersized or oversized particles. All percentage values of this example apply to percentage by mass (i.e., mass fraction [m/m]).

STAGE 1

Briefly, should equal shares of each of the coarse aggregate fractions—2/5.6, 5.6/8 and 8/11.2—be secured, the mixture presented in Table 6.6 is the result.

Even though the coarse aggregate fraction totals 75% (m/m) of the mixture and everything seems to be all right with regard to its content, the desired discontinuity of gradation is impossible to achieve by the use of this uniform distribution of constituents (at 25% each) (Figure 6.4). We can even safely say that there is a kind of continuity of gradation within the coarse aggregate fraction.

STAGE 2

Now let us divide the coarse aggregate fraction into other proportions. According to the conclusions of Example I, more coarse aggregate have to be added but material 2/5.6 mm has to be removed to break the gradation curve and pull it down at the 5.6 mm sieve. In Example II, Stage 1, 25% of the coarse particles pass through the 2-mm sieve (the fixed quantity, as assumed) and 50% through the 5.6-mm sieve (Figure 6.4). The expected level of particles larger than

---

TABLE 6.6

Example of the SMA Mixture with a Uniform Distribution of the Coarse Aggregate Fraction among Three Fractions (Example II, Stage 1)

Component of an Aggregate Mixture	Content, (m/m)	Comments
Filler	10%	Aggregates for mastic
Crushed sand 0/2	15%	
Coarse aggregate 2/5.6 mm	25%	Coarse aggregate >2 mm
Coarse aggregate 5.6/8 mm	25%	(total 75% [m/m])
Coarse aggregate 8/11.2 mm	25%	

5.6 mm exceeds 50% (e.g., 70% [passing through the sieve 5.6 mm at 30%]). The result is that the increase in material between sieves 2 and 5.6 (passing at 25% and 30%, respectively) should amount to approximately 5%, which is tantamount to the statement that screening of the fraction 2/5.6 mm in the given aggregate mix should be approximately 5%. Screening of the fraction 2/5.6 mm here is identical with the aggregate 2/5.6 mm. Therefore chippings of 2/5.6 mm have been reduced from 25 to 5%.

The only thing remaining is to decide where to add the regained 20% of the coarse aggregate—to chippings of 5.6/8 or to 8/11.2 mm?

Let us suppose that the regained 20% is being added to 5.6/8 mm chippings, leaving the fraction 8/11.2 unchanged. Table 6.7 presents the mix composition at that stage, while Figure 6.5 illustrates its gradation curve. It should be remembered that the share of material larger than 2 mm remains unchanged and still amounts to 75% (m/m).

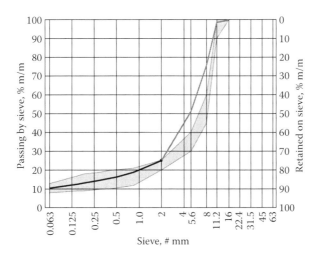

FIGURE 6.4   Example of an SMA mixture with an even distribution of the coarse aggregate fraction among three fractions (Example II, Stage 1).

TABLE 6.7

Example of the SMA Mix with an Uneven Distribution of the Coarse Aggregate Fraction among Three Fractions— Predominantly Aggregates 5.6/8 (Example II, Stage 2)

Component of Aggregate Mixture	Content, (m/m)	Comments
Filler	10%	Aggregates for mastic
Crushed sand 0/2	15%	
Coarse aggregate 2/5.6 mm	5%	Coarse aggregate >2 mm
Coarse aggregate 5.6/8 mm	45%	(total 75% [m/m])
Coarse aggregate 8/11.2 mm	25%	

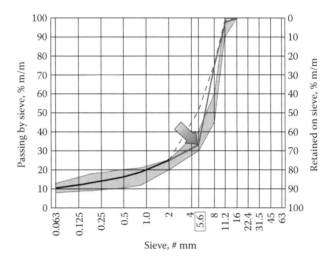

FIGURE 6.5   Example of SMA mix with an uneven distribution of the coarse aggregate fraction among three fractions—the effect of a decrease in quantity of aggregate 2/5.6 and the supplement of 20% chippings 5.6/8 (Example II, Stage 2).

A better gap-gradation is a distinctive feature of the achieved gradation curve. The broken shape of the gradation curve of that mix is clearly noticeable. Despite that improvement, the mix is still not acceptable because there is too little material larger than 8 mm. So replacing the 20% chippings 2/5.6 removed with only chippings 5.6/8 has not led to the defined goal. Therefore let us try another variant.

STAGE 3

This time, let us add that 20% removed from 2/5.6 mm chippings to 8/11.2 mm chippings (that makes 25% + 20% = 45%), leaving the content of fraction 5.6/8 the same as in Stage I (25%). The composition of the new mix is shown in Table 6.8, while its gradation curve is shown in Figure 6.6.

The achieved mix falls between the upper and lower gradation limits. Obviously it still needs some refining within the sand fraction, but at the moment the main topic is the coarse aggregate. Perfectionists would say that a bit of "messing about" with the coarse fraction could be useful, by lowering the gradation curve even more on the 8 mm sieve, for example. But the question is, is it worth it?

After all, lowering the gradation curve on the 8 mm sieve will increase the share of particles larger than 8 mm, which means a more coarse gradation. The mix with a predominant share of the fraction 8/11.2 will make a strong skeleton; however, it will be characterized by a high value of voids in mineral aggregate (VMA) that requires a high binder content to achieve a suitably low content of air voids. We also obtain better (deeper) macrotexture, which means better skid resistance with high speed measurements. The down side is that such a mixture will have considerably higher permeability. So is this worth doing?

Despite this, let us make the last correction of the mix.

TABLE 6.8

Example of the SMA Mix with an Uneven Distribution of the Coarse Aggregate Fraction among Three Fractions— Predominantly Chippings 8/11.2 (Example II, Stage 3)

Component of Aggregate Mixture	Content, (m/m)	Comments
Filler	10%	Aggregates for mastic
Crushed sand 0/2	15%	
Coarse aggregate 2/5.6 mm	5%	Coarse aggregate >2 mm
Coarse aggregate 5.6/8 mm	25%	(total 75% [m/m])
Coarse aggregate 8/11.2 mm	45%	

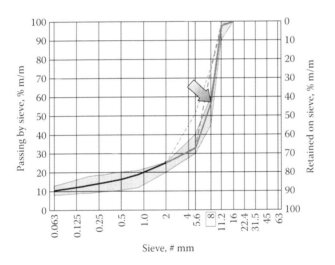

FIGURE 6.6 Example of SMA mix with an uneven distribution of the coarse aggregate fraction among three fractions—the effect of the decrease in quantity of aggregate 2/5.6 and the supplement of 20% aggregate 8/11.2 (Example II, Stage 3).

STAGE 4

After examining the gradation curve in Figure 6.6, we can conclude that lowering the curve on the 8-mm sieve simply means a reduction in passing from 55 to 45%, which is an increase in the quantity of particles bigger than 8 mm from 45 to 55%. This move means that we should increase the content of 8/11.2 by 10%. This action should be balanced, so 10% has to be taken off the aggregate 5.6/8. As a result, the achieved composition of the mix is shown in Table 6.9 and the gradation curve in Figure 6.7.

TABLE 6.9

Example of the SMA Mix with an Uneven Distribution of the Coarse Aggregate Fraction among Three Fractions—Predominantly Chippings 8/11.2 (Example II, Stage 4)

Component of Aggregate Mixture	Content, (m/m)	Comments
Filler	10%	Aggregates for mastic
Crushed sand 0/2	15%	
Coarse aggregate 2/5.6 mm	5%	Coarse aggregate >2 mm
Coarse aggregate 5.6/8 mm	15%	(total 75% [m/m])
Coarse aggregate 8/11.2 mm	55%	

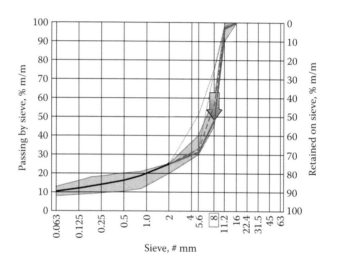

FIGURE 6.7 Example of SMA mix with an uneven distribution of the coarse aggregate fraction among three fractions—the effect of the decrease in quantity of aggregate 2/5.6 and 5.6/8 combined with adding 10% aggregate 8/11.2 (Example II, Stage 4).

6.3.2.2.1 Finalizing the Changes in the Aggregate Mix

As a result of the actions undertaken during Stages 1–4, the present mix is distinguished by a very high proportion of coarse particles (the strong skeleton) and the maximum discontinuity of gradation allowed by the gradation curves of SMA 0/11 according to ZW-SMA-2001. The share of the coarse aggregate fraction was fixed all the time at the level of 75% (m/m). Keeping in mind the impact of the size of coarse aggregate fraction on the content of voids and binder, specimens can be prepared and then tested to check their characteristics, and finally some final refinements to the fraction in question can be made.

But there is still a question, is it a good SMA mixture? Let us compare our newly designed very coarse SMA to the German recommendation and the 30–20–10 rule.

6.3.2.2.2 Comparison of Results of Example II Using German Proportions of SMA Composition

In Chapter 2, Table 2.1 cites the recommended ratios of individual SMA coarse aggregate fractions from the German DAV handbook (Drüschner and Schäfer, 2000). The comparison of the achieved result from Example II with ratios required in Germany is shown in Table 6.10.

The comparison in Table 6.10 shows that our SMA differs both from the original Zichner proportions and the contemporary ones recommended in Germany. The original German SMA does not contain such a great amount of the coarsest grains. Therefore let us design the same mix according to the German DAV proportions. The result is shown in Figure 6.8. The gradation curves of DAV and Zichner have a gentler shape, making laydown and compaction easier. Not using the maximum quantities of the coarsest grains makes the mix less open graded. Such an SMA mixture will probably be less permeable to water.

6.3.2.2.3 Comparison of Results of Example II with the 30–20–10 rule of SMA Composition

According to the rule described in Section 6.2.2, proper stone-to-stone contact is created if the percentage of aggregate passing the 0.075 mm, 2.36 mm, and 4.75 mm sieves equals 10%, 20%, and 30%, respectively. Table 6.11 shows the comparison between the achieved result of Stage 4 and ratios according to the 30–20–10 rule.

There are some noticeable differences. First, the 30–20–10 SMA should contain more aggregates larger than 2.36 mm (80%), whereas the relevant SMA of Example I was designed at only 75% on the 2.0 mm sieve. With regard to particles larger than

---

TABLE 6.10
Recommended Ratios of SMA 0/11S according to German DAV Handbook Compared with the Result of SMA Design in Example II

SMA Coarse Fraction Components	Original Zichner's Proportions of MASTIPHALT[a]	Recommended Ratios within SMA Coarse Aggregates' Fraction According to the German Guidelines for SMA 0/11S (Mass Fractions)[b]	SMA of Example II, Stage 4	Ratios of the Mix Converted According to the German DAV Guidelines (for 75% of Grains Retained on Sieve 2.0 mm)
Fraction 2/5.6	10%	1 part	5%	10.7%
Fraction 5.6/8	27%	2 parts	15%	21.4%
Fraction 8/11.2	38%	4 parts	55%	42.9%
Total	75%	7 parts	75%	75.0%

[a] See Table 6.4.
[b] From Drüschner, L., Schäfer, V., Splittmastixasphalt. DAV Leitfaden. Deutscher Asphaltverband, 2000.

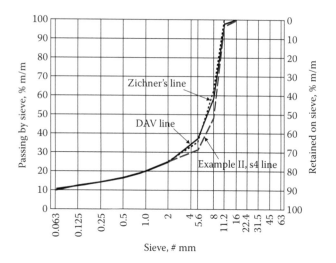

FIGURE 6.8 SMA mix of Example II, Stage 4, adjusted to the ratios recommended by the German DAV handbook compared with its original gradation and Zichner's proportions.

TABLE 6.11
Recommended SMA Ratios according to the 30–20–10 rule Compared with the Result of SMA Designed in Example II

Gradation	Recommended Approximate SMA Ratios According to 30–20–10 Rule (for SMA 0/12.5)	SMA 0/11 of the Example II, Stage 4, the Approximate Conversion of Passing to U.S. Sieves
Passing by the 0.075-mm sieve	10% (m/m)	10% (m/m)
Passing by the 2.36-mm sieve	20% (m/m)	On the 2.0-mm sieve 19% (m/m)
Passing by the 4.75-mm sieve	30% (m/m)	On the 5.6-mm sieve 31% (m/m)

4.75 mm, the result obtained in Example II is consistent with the requirement concerning 30%. Figure 6.9 shows the SMA corrected in such a way that its ratios are in conformity with the assumptions of the 30–20–10 rule.

The SMA designed according to the 30–20–10 rule is more gap graded than our SMA in Example II, which is especially evident in the percent passing the 2.0-mm sieve. In general, mixtures with such a strong gap in grading are harder to compact and are more permeable. On the other hand, one can get very strong skeleton with such a clear gap grading (stone-to-stone contact).

6.3.2.2.4 Zichner's and DAV's SMA versus 30–20–10 Rule

The 30–20–10 rule has more or less rigid proportions of gradation (expressed by amounts passing through selected sieves). Zichner proposed the amount of coarse

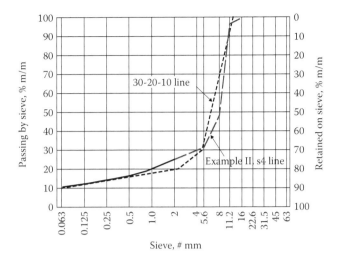

FIGURE 6.9   An SMA mix of Example II, Stage 4, compared with the line of the 30–20–10 rule.

aggregates over a very wide range (65–80%), with a recommendation for 70–75%. His MASTIPHALT (SMA 0/12.5, mm) with 75% of grains larger than 2 mm has a gentle gap gradation with also more or less fixed proportions. There is no doubt that the 30–20–10 rule creates quite different mixtures than those created according to Zichner's ideas. The other issue is with DAV's proportions (Table 6.10) because these express the proportions between aggregates inside the coarse fraction. Hence the proportions mentioned can be applied to the amount of coarse aggregates fraction within the permitted range (e.g., 70–80%). In Figure 6.8 we compared the DAV line applied with a fixed 75% of coarse aggregates as we designed in Example II. Now we can evaluate the DAV proportions applied to a new content: 80% of coarse aggregates (similar to the 30–20–10 rule). After calculation of the gradation larger than the 2.0 mm sieve, according to DAV rules we can see grading curves as shown in Figure 6.10. The final assessment of the 30–20–10 rule line is slightly below the extreme line permitted by DAV proportions at 80% of coarse aggregate fraction. As is the case in Germany, there is a trend to decrease the amount of coarse aggregate content to 73–76% (Drüschner, 2005). Taking into account this assumption, the 30–20–10 rule line is too low.

One can see in Figure 6.10 that the latest German regulations TL Asphalt-StB 07 for SMA 11S requires 35–45% of material passing a 5.6 mm sieve. It shows that Zichner's or DAV's proportions (for 70–76% of stones) are still in use.

6.3.2.2.5 Summary of Example II

The following is a list of the conclusions drawn from analyzing Example II:

- All four variants have preserved a fixed content of chippings at a level of 75%.

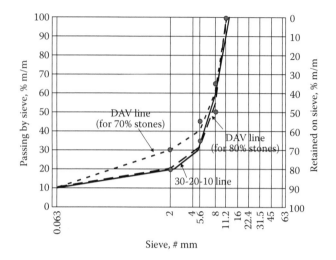

FIGURE 6.10   Comparison of SMA gradations: 30–20–10 rule (solid line) and DAV lines for 70% and 80% of coarse fraction (dotted lines). The points show gradation limits in coarse fraction established in the newest German TL Asphalt-StB 07 for SMA 11S.

- The ratios of the coarse aggregate fractions cannot be even because this causes a loss of the necessary discontinuity of gradation.
- Quantities of the finest grains in the coarse fraction should be reduced when composing the total coarse aggregate fraction.
- The content of grains larger than 2 mm (the coarse aggregate) in an SMA aggregate mix do not explicitly determine its aggregate skeleton and properties; aggregates passing the 5 or 8 mm sieves are also needed.
- Increasing the quantity of particles larger than 5 mm leads to opening the mix; this effect is even more noticeable when raising the content of particles larger than 8 mm.

6.3.2.2.6   Grading of the Coarse Aggregate Fraction versus the Distribution of Air Voids

Investigations carried out in the Netherlands (Voskuilen, 2000) have proved that the gradation within the coarse aggregate fraction exerts an impact on the distribution of voids in a mix. Briefly, the conclusions drawn in the Netherlands are as follows:

- A finer graded coarse fraction is characterized by a greater amount of smaller pores more evenly distributed through the mix, which brings about better interparticle contact and, in contrast, increases the risk of shoving grains aside by larger particles of the sand fraction.
- A coarser graded coarse fraction is characterized by a smaller amount of larger-sized pores unevenly distributed over a mix.

6.3.2.2.7 Determining the Size of Active Particles

After exercises in changing ratios within the coarse aggregate fraction, it is time to explore the question of the influence of the size of active particles on the mix. As we remember from Chapter 2, active grains are those making an aggregate structure that carries loads. The problem of actively setting up the SMA skeleton by particles of a certain fraction—say, 2/4 (or 2/5.6) mm—was discussed there. According to the German approach to SMA, that size of particle could be used for that purpose, though to a limited extent (as the German ratios suggest that in SMA 0/11 only one seventh of all coarse aggregates should be of size 2–5mm). According to the U.S. approach, this size should not be used, although that depends on the maximum size of the SMA aggregates, or NMAS. The lower limit sieve, from which active particles are counted, is called the breakpoint (BP) sieve in the United States.

The adopted classification in the United States—in NAPA SMA Guidelines QIS-122—imposes lower size limits for coarse (active) particles based on NMAS as follows:

NMAS: 25 mm	BP sieve = 4.75 mm
NMAS: 19 mm	BP sieve = 4.75 mm
NMAS: 12.5 mm	BP sieve = 4.75 mm
NMAS: 9.5 mm	BP sieve = 2.36 mm
NMAS: 4.75 mm	BP sieve = 1.18 mm

At any rate, coarse particles 2.36/4.75 mm (below 4.75) have been regarded as active ones in SMA 0/9.5 mm. In SMA 0/4.75 mm, the fraction 1.18/2.36 mm is also considered an active one (as are all larger ones). In coarser mixes, aggregates above 4.75 mm are regarded as active.

The selection of the BP sieve influences not only on the shape of the gradation curve but also the properties of SMA mixtures. Generally, the larger the BP sieve, the stronger the predominance of coarse (active) particles in a mix. One can safely say that the coarse aggregate fraction becomes more single sized as the discontinuity of gradation becomes stronger. When estimating gradation curves for various BP sieves, the conclusion can be drawn that the larger the BP sieve, the further the position of the breaking point of the gradation curve is moved to the right. And thus one can also say that the larger the BP sieve, the more open the mix and the more binder is required.

Results of some work in the United States (Cooley and Brown, 2006) justify saying that raising the size of the BP sieve results in the following consequences to the properties of an SMA mixture because it increases:

- The contents of air voids in the aggregate mix (i.e., VMA), causing a substantial rise in the optimum quantity of binder—the very high volume of air voids in the coarse aggregate skeleton must be filled in with binder
- The SMA resistance to permanent deformation, which is an advantageous effect

- Permeability of the mixture—with the same content of voids in a compacted SMA mixture, the permeability is higher with a larger BP (for more on permeability, see Chapter 12).

6.3.2.2.8   Summary of Part II on the Gradation of the Coarse Aggregate Fraction

A fixed coarse aggregate content was assumed in Part II of the design example. We explored the subdividing of the coarse aggregate fraction and its subsequent consequences. We saw that the shape of the gradation curve in the area larger than 2 mm has great significance on the properties of an SMA mixture. Thus, by increasing the predominance of very coarse grains, we increase the following:

- Resistance to permanent deformation (in general, but not in all cases)
- The content of VMA
- The binder quantity
- The permeability of a course

A final remark: setting up the skeleton using only the coarsest particles—namely, creating a single-sized mixture—will bring about possible problems with the interlocking of the skeleton grains; contrary to expectations, such a course will be of poor quality.

6.3.2.3   Part III: Shape of Particles of the Coarse Aggregate Fraction

The shape of the aggregate particles, from flat and elongated to cubical ones, exercises a certain influence on the SMA mixture. A high content of flat and elongated particles has the following effects:

- Increases the content of air voids in a compacted mixture of coarse (active) aggregates
- Decreases the workability of a mixture
- Increases the risk of fat spots appearing when compacting the SMA course

In the German DAV handbook (Drüschner and Schäfer, 2000), attention has been paid to the impact of the shape of particles from the fraction 2/5.6 mm on the content of voids in SMA, especially with reference to the SMA mixtures 0/8 and 0/8S.

6.3.2.4   Part IV: Impact of the Grains' Density

The appearance of significant differences in density among individual fractions of the aggregate mix compels us to discuss volume relations in the aggregate mix and necessary adjustments to the binder content

Substantial differences in the densities of aggregates combined in an SMA can cause numerous problems. This situation happens rather seldom; however, it is possible to find very light material combined with very heavy aggregate (e.g., densities of approximately 2.400 and 3.100 Mg/m³, respectively). When designing a particular mix, results of the sieve analysis of constituent aggregates, as well as the overall gradation curve of a mix, indicate the gradation sieve distribution in mass units. In fact, mass and volume distributions do not correspond with each other if there are substantial differences in the densities of aggregates. Hence, guidelines have been created to regulate the allowable difference between densities. For example, a difference of approximately 0.2 Mg/m³ is allowed according to AASHTO M 325-08; if it is higher, the sieve distribution should be converted into volume units. Conducting the sieve analysis and determining the results in volume units have not always been practiced outside the United States

Besides problems with the aggregate mix, the use of aggregates with different densities brings about the necessity of correcting the binder content. For that purpose, correction coefficients have been used all over the world. Approximate, or "framework," binder contents in SMAs have been detailed in various reference documents (e.g., standards, guidelines, recommendations) from many countries. The given minimum quantities of binder have been established based on a reference density of an aggregate mix.

For example, in the NAPA SMA guidelines QIS-122 the minimum content of binder in SMA amounts to 6.0% (m/m), but that is the value for a reference aggregate density equal to 2.750 Mg/m³. If the aggregate mixture density is different from the reference one, an adjustment should be made according to the following principles:

- 0.1% of the binder for each 0.05 Mg/m³ of difference between the density of an aggregate mix and the reference density (2.75) could be added or subtracted.
- For a density smaller than 2.75 Mg/m³, the correction bears the plus sign (+); for a density larger than 2.75 Mg/m³, it bears the minus sign (−).

Since 2006 the rules for correcting binder contents have been standardized in European member states of the CEN. The correction coefficient α for aggregate mixes with densities different than 2.650 Mg/m³ has been adopted in the European standard EN 13108-5 on SMA. The minimum content of binder stipulated by the requirements of this standard should be adjusted depending on α calculated as follows:

$$\alpha = \frac{2.650}{\rho_a}$$

where:

- α = The coefficient adjusting the binder content
- ρ_a = The density of the aggregate mix determined according to EN 1097-6

For a thorough description of the requirements of the EN 13108-5 standard, see Chapter 14.

6.3.2.5   Laboratory Example

The relationship between the contents of air voids and coarse grains is well-illustrated by the laboratory example described next.

Two mixes, identified by letters E and F, were produced in laboratory conditions to demonstrate the differences between SMA mixtures with the following different gradations:

- Mix E is characterized by a lesser discontinuity (more uniformity) of gradation.
- Mix F, designed according to U.S. gradation curves using NAPA SMA Guidelines QIS-122, has a much higher content of coarse aggregate particles.

Both mixes were prepared with the use of a combination of sieves. The same combination has been applied to present gradation curves and to perform analyses of the aggregate mixes. The gradation curves of aggregate mixes E and F are shown in Figure 6.11. The aggregate mixes are compared in Table 6.12.

Figure 6.12 shows photographs of cross sections of the Marshall specimens of mixes E and F.

Mix F is distinguished by a higher discontinuity of gradation (aggregate 5.6/8 is missing from the composition) and a lower content of the sand fraction (by 2.5%). It is worth observing that the difference between contents of the fraction larger than 2 mm amounts only to 1.9%. The most significant differences appear on the 6.3 and 8.0 mm sieves. Differences between the mixes increase along with an increase in the sieve size. Mix F has been made in accordance with the U.S. gradation curves using NAPA QIS-122, which is based on an assumption that the direct contact within coarse

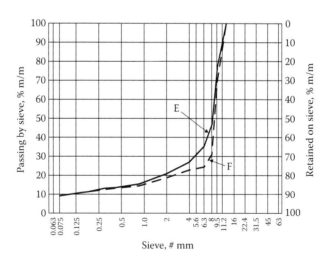

FIGURE 6.11   Grading curves of mixtures: E (solid line), F (dotted line)

TABLE 6.12

Composition of the Aggregate Mixes E and F

Composition	Mix E % (m/m)	Mix F % (m/m)	Difference Between Mixes
Properties of an Aggregate Mix			
1. Content of the filler fraction	9.1	9.7	+ 0.6
2. Content of the fine aggregate fraction	11.2	8.7	–2.5
3. Content of the fraction > 2.0 mm	79.7	81.6	+ 1.9
4. Content of the fraction > 4.0 mm	73.4	77.4	+ 4.0
5. Content of the fraction > 6.3 mm	65.0	76.0	+ 11.0
6. Content of the fraction > 8.0 mm	55.0	69.0	+ 14.0

(a) (b)

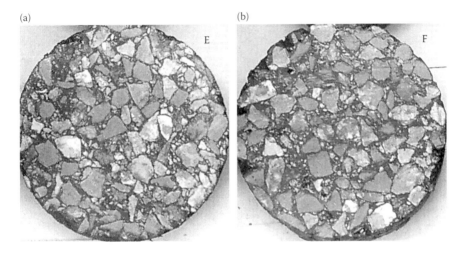

FIGURE 6.12 Photograph of cross sections of Marshall specimens of mixes E and F. (Photo courtesy of Halina Sarlińska.).

chippings should be guaranteed—that is, the condition of stone-to-stone contact has to be satisfied. Both SMA mixtures were manufactured with the same fixed amount of binder (6.4% by mass), while the differences between them are obvious when comparing the contents of voids in the Marshall specimens. The air void contents are as follows: mix E had 4.7% (v/v) and mix F had 5.2% (v/v).

Thus an increase in the content of coarse particles—and furthermore in the coarsest fraction of the coarse particles—brings about a definite opening of the SMA mixture. In other words, when moving the gradation curve toward higher contents of coarse grains, one should take into account the increase in the binder content, and probably the stabilizer as well.

6.3.3 Designing an Aggregate Mix Less than 2 mm

When designing the gradation of aggregate smaller than 2 mm (filler and fine aggregate), it should be kept in mind that the excellent properties that allow SMA to resist permanent deformation are connected mainly with a coarse aggregate skeleton. Mastic made of filler, fine aggregate, and binder should play the role of bonding and sealing the coarse aggregate, while its quantity cannot be greater than the free space left among the compacted active grains. See Chapter 7 for a discussion of the Dutch method of designing the volume of mastic in SMA.

6.3.3.1 Content of the Fine Aggregate Fraction

In classic SMA composition and in regulations introduced all over the world, the total content of grains smaller than 2 mm has generally been in the range of 15–30% (m/m). When we add the typical filler content (8–13%), we receive up to 22% from the sand fraction (0.063/2.0 mm). But when designing the content of fine aggregate in SMA, one should remember the increase in the content of fine particles during compaction due to crushing and wearing of the coarse particles.

Is the sand fraction desired in a mix? Looking at the shape of an example SMA gradation, we can imagine a mix designer adding all the permitted quantity of filler (i.e., approximately 13% [m/m]) instead of 0.063/2.0 material. This example is illustrated in Figure 6.13. As can be seen, the gradation curve stays within the limits up to the 0.85 mm sieve. Then it takes the "low route," meaning that there are too few fine particles, which are needed for mastic creation. Undoubtedly, composing an SMA without the material 0.063/2.0 is not possible.

The sand fraction is indispensable because building a mastic with only filler grains makes achievement of the expected features of a newly designed SMA impossible. So is it possible to determine the best course of a gradation curve in the area below

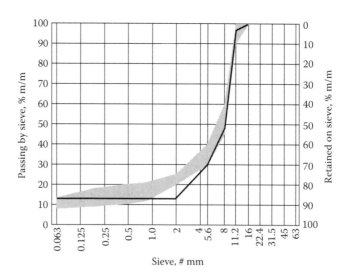

FIGURE 6.13 Example of a SMA aggregate mix gradation without fine aggregate (0/2 mm).

2 mm? There is not an unequivocal answer to this question because a lot depends on the type and properties of filler and features of the 0.063/2.0 material, too. The intended use of the designed mix is also of great significance. However, the following is worth bearing in mind:

- Guiding the curve upward enhances the risk of closing the mix and raises the threat of excessive mastic and the appearance of fat spots.
- Guiding the curve downward enhances the risk of an excessive opening of the mix.
- Designing using the maximum quantity of filler and the minimum amount of the sand fraction is disadvantageous and risky.
- Care ought to be taken so that an increase of the sand fraction can be observed on subsequent sieves to supply enough material for making mastic.
- The quantity of filler should fluctuate around the middle of the allowable range (i.e., about 9–10% [m/m]) to make possible the collection of material on sieves smaller than 1.0 mm and to prevent the gradation curve from rising upward.
- Non manufactured (natural) sand may be applied only for SMA layers on roads with low traffic volumes.

6.3.3.2   Filler Content

In the majority of worldwide regulations for SMA, the content of particles passing through the smallest sieve (0.075 or 0.063 mm) generally ranges between 8% to 13% (m/m). However, adopting extreme quantities may be a risky business—that is, 8% can lead to building too little mastic. On the other hand, a large quantity of filler (e.g., approximately 13%) may generate too high a content of mastic, making it susceptible to overstiffening or increasing the risk of forming fat spots.

It has been discussed in Chapter 3 that the optimum relationship between quantities of filler and binder is best illustrated by the filler-to-binder ratio (by weight or volume). This means that each quantity of filler corresponds to a certain optimum amount of binder. The details behind this assumption are inexact, resulting perhaps from experience with a run-of-the-mill filler in a given country. After reading Chapter 3, it should be clear that there are all sorts of fillers and the differences between them do not lie only in one specific area (e.g., gradation, degree of grinding) but also in the different content of voids in the compacted filler (determined using Rigden's method).

6.4   DESIGNING A BINDER CONTENT

Selecting the binder content in a design SMA mixture is relatively easy. With a correctly designed aggregate mix, it is enough to remember an appropriate content of voids in compacted samples. A thorough understanding of that subject will surely be made much easier by reading Chapter 7, including the description of both the U.S. and Dutch methods, and the section in Chapter 8 on preparing samples.

The majority of SMA guidelines have stipulated minimum contents of binder for a specific SMA mixture, and a limitation on the maximum quantity of binder has occasionally appeared. In each case, it should be kept in mind that these limits have been introduced in relation to the expected density of an aggregate mix (see Section 6.3.1.4).

Designing the binder content in SMA is the next stage of work after fixing the composition of an aggregate mix (using any method). Normally, the aim is to determine the content of the binder, that enables achieving the expected level of voids in compacted mix samples.

The method of compaction (Marshall versus gyratory) influences the final optimum binder content, therefore it is very important to use equivalent compactive efforts. For example, the number of rotations should equal 2×50 blows in Marshall or, alternatively, the number of gyratory revolutions should be standardized and used consistently. Improper parameters of gyratory compaction lead to misleading results of optimum binder content. The description of this topic is in Chapter 8.

Using analytical formulae that enable the determination of the optimum quantity of binder in a mixture is increasingly rare. These equations were invented based on the conversion of the specific surface area of an aggregate mix, and the determination of the film thickness needed to coat the aggregate. Nevertheless, it is necessary to say clearly that the probability of finding the optimum quantity of binder is not high because the most frequently used conversion factors were adopted for AC but not for SMA. Naturally, they do not take into account the specificity of forming voids among particles of a skeleton as we saw in Part I.

The *a priori* assumption of a specific content of binder in SMA is another very interesting aspect of selecting an optimum quantity of binder. Given an optimum binder content, an adequate aggregate mix is selected to allow the required amount of binder, making use of rules already known by the reader. The first of these relations is between the content of voids and gradation of the coarse aggregate fraction. This approach is used in the Dutch method (see Chapter 7).

6.5 REQUIREMENTS FOR AN SMA MIXTURE DESIGN

Volumetric properties are among the most frequently cited requirements for SMA mixtures checked at the laboratory level. The primary requirement is to ensure the needed content of voids in compacted samples. Mechanical requirements (e.g., stability) are seldom determined, whereas performance-related properties (e.g., resistance to rutting) can be more often seen in specifications. Table 6.13 shows a short summary of different types of requirements. The corresponding summary of requirements for SMA in accordance with the European classification system after EN 13108-5 can be found in Chapter 14.

Requirements for laboratory-designed SMA mixtures according to the European standard EN 13108-5 can be found in Chapter 14 (see Table 14.3).

Upon completing the design of an aggregate mix and the contents of binder and stabilizer (see Chapter 8), it is worth investigating whether the properties of a newly prepared SMA mixture can yield the characteristics required by the customer after construction. A more detailed description of those qualities and related research may be found in Chapters 10 and 12.

TABLE 6.13
Summary of Requirements for SMA Mixtures in Various Countries

Property	Requirement	Example of Occurrence	Comments
Air void content in compacted SMA samples	2.0%–4.0% (v/v) 3.0%–4.5% (v/v) for heavy traffic	Majority of countries	Declared recommended range
	≤5.0% (v/v)	New Zealand	Declared upper limit (related to so-called *refusal density*)
	–5.0% (v/v)	Netherlands	Recommended value for SMA 0/11 in heavy duty pavements
VMA mixture	≥17% (v/v)	United States	17% threshold used at the stage of production control; 17.5 or 18.0% (v/v) threshold recommended at the recipe stage
	≥19% (v/v); ≥20% (v/v)	South Korea	19% limit for SMA 0/10; 20% limit for SMA 0/8
Voids filled with binder (VFB)	70%–85% 80%–90%	Slovenia Finland	SMA for heavy duty traffic Recommended 85%
Air voids content in compacted SMA samples with maximum compaction energy	≥2.0% (v/v)	New Zealand	Depending on design method[a]
	≥2.5% (v/v)	Czech Republic [b]	Determined when compacting Marshall samples with the effort 2×100
Marshall stability	≥6.0 kN	Czech Republic [b]	Only when applying Marshall method for designing SMA
Uniaxial creep stiffness modulus (static mode)	≥16 MPa	Poland [b]	Used before implementation of WTT apparatus
Wheel tracking test	≤5%	Poland [b] Austria [b]	EN 12697-22 method— small device; temperature 60°C; 10,000 cycles, method B (PRD_{AIR} or WTS_{AIR})

(Continued)

TABLE 6.13 (CONTINUED)
Summary of Requirements for SMA Mixtures in Various Countries

Property	Requirement	Example of Occurrence	Comments
Resistance to water (ITSR)	≥ 70%	United States	AASHTO T283 method (see Chapter 12)
	≥ 80%	Slovakia[b]	EN 12697-12 method (see Chapter 12)
	≥ 90%	Poland[b]	
Binder/mastic drainage	≤ 0.3% (m/m)	The majority of countries	Testing with Schellenberg's method or similar

Note: AASHTO = American Association of State Highway and Transportation Officials; PRD = Proportional Rut Depth, result of WTT; SMA = stone matrix asphalt; VMA = voids in mineral aggregate; WTS = wheel tracking slope, result of WTT; WTT = wheel tracking test.

[a] If samples are prepared with gyratory compactor, the maximum density is read after 300 or 350 gyrations.

[b] Since 2006 requirements in European CEN members have to be established according to EN 13108-5 and test methods in EN 12697 with test conditions as in EN 13108-20.

6.6   SUMMARY

- SMA mixtures may be used both in wearing and intermediate layers.
- The suggested minimum thickness of a course equals 3.5–4 times the maximum aggregate size.
- As a general rule, heavy traffic loadings require coarser mixtures. In these cases the mixtures 0/11 and 0/12.5 mm are the most popular solutions. Unfortunately, such mixtures also have weak points, including low noise reduction, higher permeability, and worse antiskidding properties in comparison with finer mixtures.
- The coarse aggregate fraction
  - When one designs the composition of the coarse aggregate fraction, to achieve the best gap-gradation the percentages of the finest and intermediate fractions should be reduced but the proportion of the coarsest ones should be increased.
  - An increase in the content of air voids in the aggregate mix and the amount of binder in SMA result from an increase in the coarse aggregate content. Specifying the content of particles larger than 2 mm in an SMA aggregate mix does not explicitly determine either its aggregate structure or its properties; it is necessary to supply information on the amount of particles larger than 5 or 8 mm (or similar sieves).
  - Increasing the share of particles larger than 5 mm leads to opening the mix; that effect is even more obvious when increasing the content of particles larger than 8 mm. Therefore, manipulating the content of the

coarsest grains offsets the strongest impact on changes in the content of air voids within the coarse aggregate fraction.

- Designing SMA with a very high content of the coarsest particles brings about the necessity of adding a larger amount of binder, and possibly more stabilizer too. Such mixtures are also characterized by higher permeability and greater compaction resistance.
- Increasing the quantity of flat and elongated particles in a mixture has the following effects:
    - Increases the content of air voids in an aggregate mix
    - Diminishes the workability of the mix
    - Increases the risk of crushing the flat and elongated particles during compaction (followed by squeezing mastic out)
- The sand fraction and filler
    - Designing an SMA using the maximum quantity of filler and the minimum amount of fine aggregate is disadvantageous
    - The quantity of filler should generally be near the middle of the allowable range, which means about 9–10% (m/m), to enable an appropriate amount of 0.063/2-mm material on sieves less than 1.0 mm.
    - Using high quantities of natural (non-crushed) sand should be avoided, and for SMAs created for heavy traffic, its use should be generally excluded.
    - A surplus of mastic in comparison with the void space among chippings causes the appearance of fat spots and a local decrease in antiskid properties.
    - Too low a quantity of mastic means a too large an air void content in a compacted course, high absorption and water permeability, and consequently a shorter life.
- The binder content
    - Corrective coefficients of the binder content that are dependent on the aggregate density should be used.
    - The content of air voids in an SMA mixture design should not be adjusted by changing the binder content; it should be done with corrections of contents and gradation of the aggregate fractions, including the following:
        - The content of the coarse aggregate fraction (see Section 6.3.1.1)
        - The ratios of constituents within the coarse aggregate fraction (see Section 6.3.1.2.)
        - Filler content
        - Binder content, as a final resort
- The content of VMA can be evidence of problems with air voids in compacted SMA specimens; an increase in VMA should be achieved by adding coarser chippings (more material retained on a 4 mm or 5 mm sieve) or by decreasing the amount of filler, while a decrease in VMA should be achieved by adding finer chippings.
- When comparing volume requirements of various guidelines, one should keep in mind major differences in procedures for determining density, which eventually change the range of results.

7 Overview of SMA Design Methods

The basic and universal rules of stone matrix asphalt (SMA) design were described in the previous chapter. Chapter 7 provides an overview of SMA design methods developed in various countries. Undoubtedly, there are many of them, so their description could be the subject of a separate book. We will focus here on the most distinctive or the most interesting ones available in the technical literature.

The literature about SMA design methods can be both instructive and creative. You may judge for yourself which method most closely fits your needs or seems to have the most merit.

7.1 GERMAN METHOD

7.1.1 DESCRIPTION OF THE METHOD

The German method is based on long-standing experience in the application of repeatable materials and mixes. Such an approach not only makes analyzing cases of successful and unsuccessful SMA much easier but also drawing conclusions and ultimately proposing changes to technical specifications.

It was discussed earlier that the recommended ratios of SMA ingredients (see Table 2.1), combined with precisely determined gradations of each aggregate fraction supplied by quarries, enable SMA design in principle almost without the use of boundary gradation curves. Obviously, such gradation curves are being published—the first one for SMA was ZTVbit-StB 84—and then widely applied in practice. The new ZTV Asphalt-StB 07 and TL Asphalt-StB 07 standards have been in use since 1st January 2009.

The following stages may be identified in the German method:

- Design composition of an aggregate mix according to gradation limits
- Determination of a series of binder contents in the mixture
- Preparation of Marshall samples (2 × 50 blows) for each variant of SMA mixture
- Determination of the volumetric parameters of the SMA specimens
- Selection of an optimum variant of the mixture meeting requirements
  - Air voids in compacted asphalt samples at 2.5–3.0% (v/v)*
  - Voids filled with binder

---

* These values are of TL-Asphalt 07 and come from the new methods of density measurements (according to EN); in the old ZTV StB 01 corresponding values were 3–4% (v/v).

- Draindown testing with Schellenberg's method
- Wheel-tracking (rutting) test (for selected types of SMA)

The content of the coarse aggregate fraction specified in ZTV Asphalt-StB 07 amounts to 70–80% (m/m) for SMA 8S and SMA 11S but only 60–70% (m/m) for SMA 5N.*

7.1.2 Volumetric Parameters

The volumetric relationship in an asphalt mixture is shown schematically in Figure 7.1 in German terms. Equations are based on TP A-08 2007 and Hutschenreuther and Woerner (2000).

Symbols in Figure 7.1 include the following:

H_{bit} = Volume of voids in compacted asphalt samples, % (v/v)
B_v = Binder volume, % (v/v)
M_v = Mineral aggregate volume, % (v/v)
$H_{M,bit}$ = Voids in mineral aggregate (VMA), % (v/v)
Obviously, the sum of all the parts should equal 100%.

$$M_V + B_V + H_{bit} = 100\%$$

The way of calculating volume parameters and defining them is outlined here.

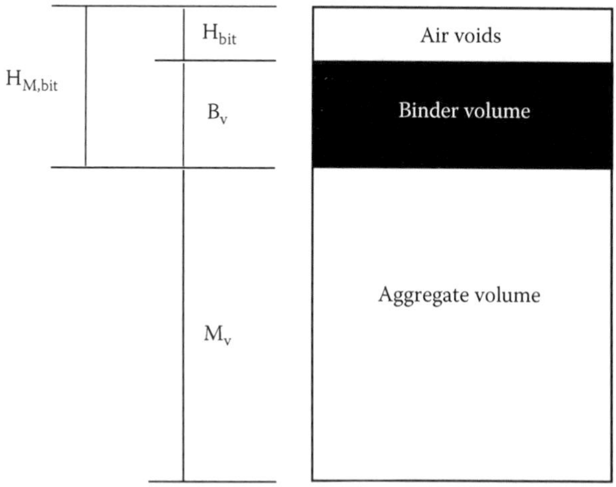

H_{bit}	Air voids
B_v	Binder volume
M_v	Aggregate volume

$H_{M,bit}$

FIGURE 7.1 Volume relationship in an asphalt mixture according to terminology adopted in Germany. (From Graf, K., Splittmastixasphalt—Anwendung und Bewährung. Rettenmaier Seminar eSeMA'06. Zakopane [Poland], 2006. With permission.)

* Designations used in German guidelines for SMA: N = low and medium traffic, S = heavy traffic (e.g., SMA 11S).

The binder volume in an asphalt mixture

$$B_V = \frac{\rho_A \cdot B}{\rho_B}$$

B = Binder content in the mixture, % (m/m)
ρ_A = Bulk density of the asphalt mixture (sample), g/cm³
ρ_B = Binder density at a temperature of 25°C, g/cm³

Volume of mineral aggregate

$$M_V = \frac{\rho_A \cdot (100 - B)}{\rho_{R,M}}$$

ρ_A = Bulk density of the asphalt mixture (sample), g/cm³
B = Binder content in the mixture, % (m/m)
$\rho_{R,M}$ = Density of the aggregate mix, g/cm³

Content of air voids in compacted asphalt mixture

$$H_{bit} = \frac{\rho_{R,bit} - \rho_A}{\rho_{R,bit}} \cdot 100\%$$

ρ_A = Bulk density of the asphalt mixture (in German *Raumdichte*), g/cm³
$\rho_{R,bit}$ = Maximum density of the asphalt mixture (in German *Rohdichte*), g/cm³

Air voids in compacted mineral aggregate (the hypothetical voids content)

$$H_{M,bit} = H_{bit} + B_V$$

B_V = Binder volume, % (v/v)
H_{bit} = Air voids in the compacted asphalt mixture, % (v/v)

Voids filled with a binder

$$HFB = \frac{B_V}{H_{M,bit}} \cdot 100\%$$

B_V = Binder volume, % (v/v)
$H_{M,bit}$ = VMA, % (v/v)

7.1.3 COMMENTS

- In most cases of designing an SMA, the steps described in Section 7.1.1 are sufficient.
- The void contents in compacted Marshall specimens after the application of various compactive efforts (2×50 and 2×75 blows [Graf, 2006]) have also been compared. For a designer, a large difference between samples of the same SMA mixture under different compactive efforts is an indicator of too high a compactability under the influence of an excessive effort. However, that practice has not been formalized.
- Some German engineers check the voids in mineral aggregate ($H_{M,bit}$) to find out if it is higher than 18% (v/v), as do their fellow U.S. engineers. However, this practice remains unsanctioned.

7.2 U.S. METHOD

This method emerged in 1990 after a tour of Europe when some U.S. engineers learned of the benefits of SMA. A series of research efforts started soon after to develop a method of designing SMA. This resulted, among other things, in publications (Brown and Haddock, 1997; Brown and Mallick, 1994) that tried to reach to the heart of the matter of SMA mixtures and suitable methods of designing and testing them.

The essential aspect of designing an SMA aggregate mix using the U.S. method is the introduction of the idea of stone-to-stone contact, or a direct contact among coarse particles. Those grains, called *active* grains, make a strong mineral matrix and give the SMA its deformation resistance. The method of testing the stone-to-stone contact has also been defined. It is called the dry-rodded test and will be explained later in greater detail. Designing SMA in the United States has been described in different publications (e.g., in NAPA SMA Guidelines QIS 122 and the standards AASHTO M325 and AASHTO R46). The method described in these guidelines will be discussed here. It consists of the following stages:

- Selecting an aggregate
- Designing a gradation curve that secures the desired interparticle contact (stone-to-stone contact)
- Selecting the gradation corresponding with the criterion of a minimum of air voids in an aggregate mix (minimum VMA)
- Selecting an amount of binder for a target content of air voids in compacted specimens of the asphalt mixture
- Checking for draindown and water susceptibility

Next we will follow the U.S. cycle of design through its successive stages.

7.2.1 STAGE 1: SELECTING AN AGGREGATE

Requirements for aggregates to be used in SMA are catalogued in Chapter 5. When differences among densities of aggregates used for composing an SMA aggregate mix exceed 0.2 g/cm³, the composition of an aggregate mix should be converted from mass into volume, and only such values can be compared with gradation limits (the requirement using AASHTO MP 8-00, currently M325).

7.2.2 STAGE 2: SELECTING A GRADATION CURVE

Composing an adequate aggregate mix is the crucial step in the design process. Adequate in this case means meeting the following conditions:

- A design gradation curve lying between the gradation limits
- Suitable contact among coarse particles—that is, the fine aggregate and filler do not interfere with the contact among the largest particles (the stone-to-stone contact is guaranteed)

When the standard method is used, at least three trial aggregate mix compositions are designed, with their gradation curves lying close to the upper, middle, and lower gradation boundaries of the allowable ranges of gradation. Obtaining three such curves involves changing ratios between the fine and coarse aggregate contents.

The quantity of filler is generally assumed to be constant, depending on the size of the SMA's biggest particle. With that in mind, the content of particles smaller than 0.075 mm should amount to approximately 14% (m/m) in the finest SMA 0/4.75 mm, while in coarser SMA mixtures the filler content should be approximately 10% (m/m). With the filler amount essentially fixed, the ratio between coarse and fine particles may be changed to adjust the position of the SMA aggregate mix gradation curve.

How do we secure contact between the coarse particles? Before discussing this, the aggregate mix should be remembered. The volume division between coarse (skeleton, active) particles and fine ones (filling, passive) is displayed in Figure 7.2. As shown, this division displays a strong particle skeleton made up of appropriate coarse grains. The term *coarse aggregates* has been intentionally omitted because,

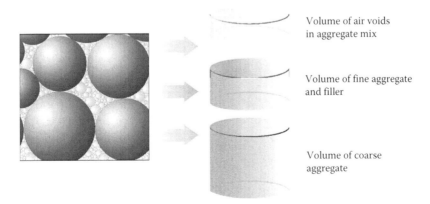

Volume of air voids in aggregate mix

Volume of fine aggregate and filler

Volume of coarse aggregate

FIGURE 7.2 The volume distribution of the elements in a mineral mix.

after all, the particles are bigger than 2.36* mm; however, these are too small to create a strong skeleton. The boundary sieves for SMA mixtures (called breakpoint [BP] sieves), from which the coarse skeletons start, depend on the following nominal maximum aggregate size (NMAS†) of the mixture:

- NMAS ≥ 12.5 mm   BP sieve: 4.75 mm
- NMAS = 9.5 mm   BP sieve: 2.36 mm
- NMAS = 4.75 mm   BP sieve: 1.18 mm

The most common mix is probably SMA 0/12.5 mm with the 4.75 mm BP sieve. According to the method, the skeleton making aggregates are 4.75 mm or larger. So these particles are not simply chippings but a slightly coarser aggregate. A 0/9.5 mm SMA with the 2.36 mm BP sieve has a skeleton made up of typical coarse fraction only (i.e. larger than 2.36 mm).

Now let us look again at an SMA 0/12.5 mm. Coarse grains making a skeleton have to be in contact with each other. Next let us consider a compacted layer consisting only of coarse aggregate particles. During compaction, the particles will become interlocked tightly so that they will come to rest against each other; and there will be nothing to prevent them from touching. As a result, we have the full, 100% stone-to-stone contact we are aiming at. Now, looking at that compacted layer of coarse aggregate, we can easily see some free space among the coarse particles. If we are able to insert passive (filling) particles into that space, then our aim of preventing the coarse particles from being shoved aside will be achieved. Putting it in a nutshell, particles smaller than 4.75 mm cannot have a higher volume than the remaining air voids in the compacted skeleton part. This way of packing the mix is displayed in Figure 7.3.

Further steps are self-evident; because all particles bigger than 4.75 mm create the aggregate structure, they have to be examined separately from the aggregate

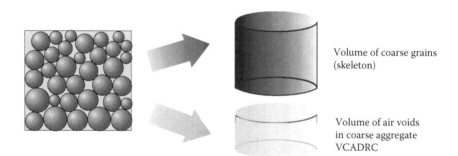

Volume of coarse grains (skeleton)

Volume of air voids in coarse aggregate VCADRC

FIGURE 7.3   The compacted skeleton part of an aggregate mix.

---

* 2.36 mm in the United States, 2.0 mm in Europe.
† NMAS stands for *n*ominal *m*aximum *a*ggregate *s*ize—a sieve one size larger than the first sieve retaining more than 10% of the aggregate.

mix. The next objective will be determining the air voids in a compacted coarse aggregate—namely, the space for filling aggregates.

7.2.3 Stage 3: Determining Air Voids in a Compacted Coarse Aggregate

The Americans have undertaken analyses on the usefulness of various methods of testing for contact between coarse particles. Ultimately, they have settled on the dry-rodded test according to AASHTO T 19-00 (Brown and Haddock, 1997). It has also been standardized as ASTM C29-97.

Let us remember what we are considering now; we are looking for the content of air voids among the compacted coarse particles that make a skeleton. Thus we are screening the coarse aggregate (regarded as the active fraction) of each of three trial aggregate mixes (three design gradation curves of Stage 2) through the boundary sieve (BP sieve) selected in accordance with the NMAS. Furthermore, three such screened samples of the coarse aggregate will be tested according to the dry-rodded method.

What does the dry-rodded method involve? All in all, it consists of compacting the coarse aggregate and determining the air voids among the particles. As a result, dry-rodded testing provides the percentage of air voids in a compacted skeleton of coarse aggregate denoted as VCA_{DRC}. It should be remembered that Volume of coarse Aggregate–Dry Rodded condition (VCA_{DRC}) has been determined for the part of an aggregate mix that is larger than the BP sieve for the size of SMA being designed.

And now the first stage of control in creating the skeleton is behind us.

7.2.3.1 Dry-Rodded Method

The dry-rodded method has been standardized in AASHTO T 19-00, where its thorough description has been included. It is recommended to perform two tests per sample and use the average value. The following gives a short outline of the equipment used and the modus operandi.

The equipment needed includes a balance, a steel tamping rod (rammer), a cylindrical metal measure, a shovel, glass calibration plate (Figure 7.4), and grease or thick glycerin. The sample of aggregate is dried in an oven to a constant mass. The sample should be about 125–200% of an amount that fits in the container. The cylindrical measure is calibrated by determining the volume using water and the glass plate (Figures 7.5 and 7.6); water-density corrective coefficients in relation to the temperature should be taken into account.

The test is performed as follows:

1. Fill the container with aggregate up to one third of its height, level the surface of the poured aggregate using your fingers, and then tamp the layer down with 25 strokes of the tamping rod, taking care to evenly distribute the strokes over the surface and avoiding hitting the bottom of the container (Figure 7.7).

2. Having completed the tamping of the first layer of aggregate, fill the container with a second layer of aggregate—this time up to two thirds of its height—and repeat the tamping procedure.

FIGURE 7.4   Prepared equipment for the dry-rodded test: the cylindrical metal measure, the steel tamping rod, and the glass plate. (Photo courtesy of Karol Kowalski and Adam Rudy, Purdue University.)

3. Having completed tamping the second layer of aggregate, fill the container with aggregate to overflowing and continue tamping down as previously described.
4. Even out the aggregate using your fingers or scrape away any excess aggregate with a rod so that protruding coarse particles will compensate for any gaps between them (Figure 7.8).
5. Determine the mass of the compacted aggregate by weighing the measure with aggregate and weighing it empty.
6. Calculations

• Calculate the bulk density of an aggregate according to the formula

$$M = \frac{G - T}{V}$$

M = Bulk density of the coarse aggregate, kg/m^3
G = Mass of a cylindrical measure and aggregate, kg
T = Mass of a cylindrical measure, kg
V = Volume of a cylindrical measure, m^3

• Calculate the void content in a compacted aggregate according to the formula

$$VCA_{DRC} = \frac{\left(G_{ca} \cdot \gamma_w\right) - M}{G_{ca} \cdot \gamma_w} \cdot 100\% \left(v / v\right)$$

FIGURE 7.5 Calibrating the measure—determining the water temperature. (Photo courtesy of Karol Kowalski and Adam Rudy, Purdue University.)

FIGURE 7.6 Calibrating the measure—determining the measure volume. (Photo courtesy of Karol Kowalski and Adam Rudy, Purdue University.)

FIGURE 7.7   Tamping down the first layer of a poured aggregate. (Photo courtesy of Karol Kowalski and Adam Rudy, Purdue University.)

FIGURE 7.8   Clearing away the excess aggregate. (Photo courtesy of Karol Kowalski and Adam Rudy, Purdue University.)

M = Bulk density of a coarse aggregate in the dry-rodded condition, kg/m^3

G_{ca} = Bulk specific gravity (dry basis) of a coarse aggregate according to AASHTO T85 = $G_{sb}{}^D$.

γ_w = Density of water, kg/m^3

7.2.4　Stage 4: Determining an Initial Content of Binder

The minimum content of binder in an SMA has been fixed at 6.0% (m/m), but a slightly higher quantity of binder in a mixture is advised. This is intended to provide protection from exceeding the lower production limit when producing the SMA.

To begin, an initial trial quantity of binder in mixtures should be adopted (the same for each of the three mixtures). The trial binder quantity should be adjusted depending on the density of the aggregate mixes. The reference density of an aggregate mix is typically 2.75 g/cm^3; if the density differs, the quantity of binder should be adjusted in accordance with this rule: each change of density by 0.05 g/cm^3 corresponds to an adjustment of binder by 0.1% (m/m). For densities less than 2.75, and for densities greater than 2.75, the adjustments are positive (+) or negative (−), respectively.

A series of samples should be made up of 12 specimens of the SMA asphalt mixture, four for each SMA trial design gradation. Then nine of the 12 samples will be compacted according to the selected method (three for each SMA design), while the remaining three samples will be used to determine the maximum density according to AASHTO T209.

Samples can be compacted using either the Marshall method or the Superpave gyratory compactor. Compaction parameters are as follows:

- Marshall hammer: 50 blows on each side of a sample 100 mm in diameter
- Superpave gyratory compactor: 100 revolutions* on samples 150 mm in diameter

Due to the risk of excessive crushing, higher compaction efforts are not recommended. The temperature of compaction samples should be determined using AASHTO T 245, which specifies that the compaction temperature is that at which the binder viscosity equals 280 ± 30 cSt, or that provided by the manufacturer of a modified binder (when applicable).

7.2.5　Stage 5: Testing an Aggregate Mix and an Asphalt Mixture

So far, we have carried out a series of tests. Let us sum up all the data at our disposal as follows:

- Air voids in the compacted coarse aggregate VCA$_{DRC}$
- Initial (or adjusted if needed) content of binder

---

* In earlier research, other numbers of rotations have been proposed based on Los Angeles (LA) abrasion loss of coarse aggregate—that is, for LA less than 30%, 100 SHRP gyratory compactor (SGC) rotations can be used, and for LA greater than 30%, 70 rotations (Brown and Cooley, 1999).

We have also made a series of compacted SMA samples. Now it is time to determine the following features for them:

- Bulk density of the compacted SMA sample
- Maximum density of the SMA mix
- Content of air voids in the compacted SMA samples (Va)
- Content of VMA
- Content of air voids in the coarse aggregate of the aggregate mix (VCA_{MIX})

Among these features, only VCA_{MIX} is a new one. It is vitally important to understand the differences between the terms VCA_{DRC}, VMA, and VCA_{MIX}. All of them describe voids in an aggregate mix or one portion of the aggregate mix. Figures 7.9 through 7.11 show their graphic representations. The following is a brief description of the three:

- VCA_{DRC}—content of air voids in the compacted coarse aggregate (coarse aggregate portion of the total aggregate mix, retained on the BP sieve)

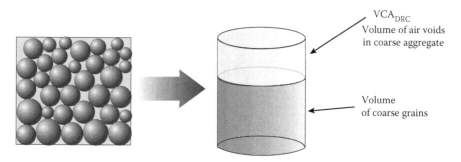

FIGURE 7.9 Definition of VCA_{DRC}—the content of air voids in a compacted coarse aggregate.

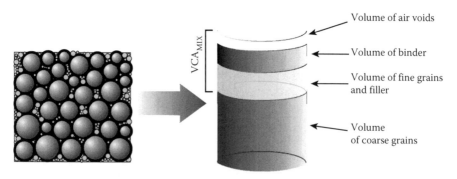

FIGURE 7.10 Definition of VCA_{MIX}—the content of air voids in the coarse aggregate of a compacted SMA mixture.

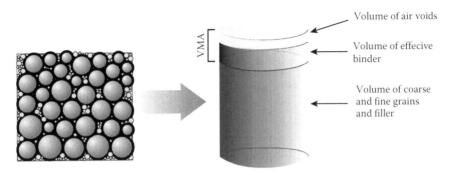

Volume of air voids

Volume of effecive binder

Volume of coarse and fine grains and filler

VMA

FIGURE 7.11   Definition of VMA—the content of air voids in the aggregate of a compacted SMA mixture.

- VCA_{MIX}—content of air voids in the coarse aggregate of a compacted SMA mixture—that is, the volume of everything but the coarse aggregate in the SMA mixture, or the volume of binder, filler, fine aggregate, and air voids)
- VMA—content of air voids in an SMA aggregate mix, equal to the sum of air void volume in a compacted SMA plus the volume of effective binder, excluding filler and fine aggregate

7.2.6   STAGE 6: SELECTING AN AGGREGATE MIX

An optimum variant should be selected out of the three defined trial design gradation curves. The selection criteria of the optimal gradation curve should consist of the smallest amount of coarse aggregate combined with the following two conditions at the same time:

- VCA_{MIX} is lower than VCA_{DRC} or the VCA ratio (VCA_{MIX}/VCA_{DRC}) is less than 1.0.
- VMA is higher than 17.0% (v/v) (usually minimum values of 17.5–18.0%).

If the VCA_{MIX} is higher than the VCA_{DRC}, the creation of a skeleton is not guaranteed. It can be changed by increasing the amount of particles bigger than the BP sieve—that is, increasing the amount of aggregates retained on the BP sieve should increase the VCA_{DRC}. So sometimes additional trial gradations must be analyzed before finding an optimal solution.

Using this method, only one mix is eventually left. The monitoring of the skeleton and aggregate structure is behind us. Now it is time for the binder.

7.2.7   STAGE 7: SELECTING AN OPTIMUM CONTENT OF BINDER

After the selection of an optimum aggregate mix, the amount of binder should be selected in such a manner that the desired content of air voids in a compacted SMA

TABLE 7.1
Requirements for a Laboratory-Compacted SMA Binder Mixture Using a Marshall Hammer or Superpave Gyratory Compactor[a]

Property	Requirement	Notes
Minimum content of binder, % (m/m)	≥6.0	Before adjustment of aggregate density
Content of air voids in a compacted sample, % (v/v)	≥4.0	Usually 3.5–4.0
Voids in a mineral aggregate VMA, % (v/v)	≥17.0	Quantity required at production in an asphalt plant; in fact it should be slightly higher in a laboratory, min. 17.5%–18.0%
VCA_{MIX}	$<VCA_{DRC}$	Appropriate filling volume among coarse particles and stone-to-stone contact are guaranteed
Marshall stability, kN	≥6.2	Suggested, not required, quantity applies only to samples compacted with a Marshall hammer
Water resistance, TSR, %	≥70	Test after AASHTO T283 (see Chapter 12)
Drain-off at the production temperature in an asphalt plant, % (m/m)	≤0.3	Test after AASHTO T305 (see Chapter 8)

Note: TSR = Tensile Strength Ratio
[a] Based on NAPA SMA Guidelines QIS 122.

sample is available. To achieve this objective, a series of samples with different amounts of binder (usually three points) should be produced, guided by the results achieved in the previous tests with an initially assumed quantity of binder. In reality, 12 samples should be made—four for each point of binder content. Having determined the bulk density and maximum density and calculated the air void content, a target quantity of binder should be selected so that the content of air voids is in the 3.5–4.0% (v/v) range.

The final amount of binder is selected based on test results. The selected SMA composition should be subjected to further testing. The final requirements are demonstrated in Table 7.1.

7.3 CZECH METHOD

7.3.1 GENERAL PRINCIPLES

The Czech method has been employed in the Czech Republic based on the Czech guidelines TP 109.

The distinctive feature of this method is the consideration of the influence of the coarse aggregate content on asphalt mixture properties. Some knowledge of the impact of coarse particle quantity in creating an SMA aggregate structure has been adopted during design. In the Czech method, particles bigger than 4 mm (called

HDK here as designated in Czech method) are regarded as coarse aggregate, making an active part of the aggregate mix.

7.3.2 DESIGN METHOD

The SMA design procedure consists of the following stages:

- Selection of the design aggregate mix using an analysis of the impact of the coarse aggregate content on SMA properties
- Determination of the optimum design content of the binder for the selected gradation

Step by step, the design proceeds as follows (for SMA 0/11).

A. Design an aggregate mix

1. Determine the properties of raw materials.
   1.1 Gradation of aggregates
   1.2 Penetration at 25°C and softening point (R&B) of the binder
   1.3 Establishing the compaction temperature for preparing samples (adjusted to the type of binder)
2. Design an aggregate mix gradation according to the required gradation limits; using this method the aggregate mix No. 3 (referred to later as mix 3) is evolving.*
3. (Based on experience)† we arbitrarily accept an optimal binder content for mix 3.
4. At this point, mix 3 has an optimal binder content (temporary); next examine the influence of changes to the aggregate mix on SMA features.
5. Design four new variants of an SMA aggregate mix in the following way.
   5.1 The binder and filler contents remain unchanged.
   5.2 Design four new aggregate mixes.
       5.2.1 Mix 1—decrease the content of HDK by −5.0% to −7.0%, and increase the content of fine aggregate by + 5.0 to + 7.0%
       5.2.2 Mix 2—decrease the content of HDK by −2.5% to −3.5%, and increase the content of fine aggregate by + 2.5% to + 3.5%
       5.2.3 Mix 4—increase the content of HDK by + 2.5% to + 3.5%, and decrease the content of fine aggregate by −2.5% to −3.5%
       5.2.4 Mix 5—increase the content of HDK by + 5.0% to + 7.0%, and decrease the content of fine aggregate by −5.0% to −7.0%

---

* We start with number 3 of the first mixture, because the next mixtures will have a different composition based on HDK content (mixes 1 and 2 with decreased HDK content, mixes 4 and 5 with increased HDK content); so the number 3 is virtually in the middle of the series.

† It is being adopted based on the experience of a process engineer (e.g., a typical amount for a given gradation). The term *optimum* is a little bit exaggerated because the truly optimal content will be described later in the text.

- With changes to the content of HDK fraction (≥ 4 mm), appropriately decrease or increase the content of the fine fraction (0.09/4 mm); the filler remains unchanged.
- While changing the quantities of HDK and fine fraction, the internal proportions of fractions (e.g., 4/8 and 8/11) should probably be maintained at a constant level.

5.3 Produce four Marshall samples of each mix (1 through 5), each containing the same quantity of binder that was adopted for mix 3 as optimal.

5.4 For each mix, determine the following:

    5.4.1 Stability according to Marshall

    5.4.2 The binder volume

    5.4.3 The content of air voids in compacted 2×50 samples (M)

    5.4.4 The content of air voids in the aggregate mix (M_k)

    5.4.5 The voids filled with binder (S_v)

5.5 Draw graphs of relationships between the elements in Step 5.4 and the content of coarse aggregate HDK.

5.6 Analyze the parameters of the mixes (1 through 5) and select the best one, based on:

    5.6.1 The analysis of the inflection point at the relationship between the content of air voids and the content of HDK in SMA samples

    5.6.2 The designer's experience

5.7 Based on the results of the analysis of the previous item, select the best gradation curve or determine a new one (i.e., mix 6).

B. Design an optimum binder content

5.8 Based on the results from Step 5.4, the optimum binder content for the selected gradation of 5.7 may be determined by producing a series of Marshall samples again, with a binder content 0.3% (m/m) higher and lower than the amount initially adopted as optimal for mix 3.

5.9 Select an optimal variant of the binder content based on the following:

    5.9.1. Stability according to Marshall greater than or equal to 6 kN

    5.9.2. Binder volume

        Greater than or equal to 14.5% (v/v) for SMA 0/11

        Greater than or equal to 15.0% (v/v) for SMA 0/8

    5.9.3. The content of air voids in compacted 2×50 (M) SMA samples, which should be from 3.0 up to 4.5% (v/v)

5.10 Conduct additional tests for the optimum content of binder in SMA.

    5.10.1. Air void content in SMA samples compacted with an excessive effort of 2×100, minimum required greater than or equal to 2.5% (v/v)

    5.10.2. Resistance for rutting, 10,000 cycles at a temperature of 50°C, required maximum less than 1.6 mm

5.10.3 Test draindown with Schellenberg's method, required to be less than 0.3% (m/m)

5.10.4 If requirements are satisfied, design is completed.

7.3.3 SUMMARY

To sum up the Czech method, it is worth noting that, despite leaving the simple basic principles of design, it facilitates examining the influence of the coarse fraction on the properties of an SMA mixture. Generally it takes into consideration all of the essential rules and relationships explained in greater detail in Chapter 6.

7.4 DUTCH METHOD

There is a widespread belief among many engineers that SMA, due to its peculiarity, should be designed by volume. The volume concept also forms the basis of an experimental method of design applied in the Netherlands.

To put it concisely, we can repeat what has been explained in the previous chapters of this book as follows:

- Air voids remain in the stone skeleton after its compaction.
- The volume of mastic, including the fine aggregate, filler, binder, and stabilizer (drainage inhibitor), has to be put into that free space.

The following description of this SMA design method includes the guidelines of 2004 and has been prepared based on information from two publications (Jacobs and Voskuilen, 2004; Voskuilen, 2000). This method was revised (simplified) in 2007; a description of the changes made is explained in Section 7.4.5.

7.4.1 GENERAL PRINCIPLES AND SOME THEORY

The principles of designing an aggregate mix and then the content of binder are presented in Figure 7.12. It is an illustration of a telescopic* method of creating SMA, which involves inserting consecutive elements into free space (air voids) in a compacted component of a larger size. In other words:

- A volume of fine aggregate is inserted into the air voids in the compacted coarse aggregate skeleton with the effect of increasing the air voids among the coarse aggregates (enlarging effect).
- A volume of filler particles is inserted into the air voids in the compacted fine aggregate.
- A volume of binder is introduced into the air voids in the compacted filler.
- The free space remaining after inserting all these elements produces the content of air voids in a compacted SMA.

---

* The *telescope* principle is a term used in a paper by Lees (1969) as the layout principle for grains in a mixture in which the finer particles fill in the remaining voids among coarser ones.

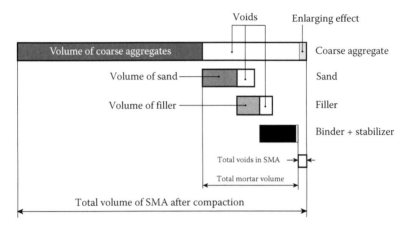

FIGURE 7.12   Placing consecutive elements of SMA in a mixture. (From Voskuilen, J.L.M., Ideas for a volumetric mix design method for Stone Mastic Asphalt. Proceedings of the 6th International Conference Durable and Safe Road Pavements, Kielce [Poland], 2000. With permission.)

Filling the air voids with the subsequent elements has already been partially demonstrated and discussed when explaining the concept of air voids in a filler (see Chapter 3).

Determining the density of all the SMA components is the basis of the design activity because the Dutch method is a volumetric-type method.

7.4.1.1   Coarse-Aggregate Skeleton

The first step, as in the U.S. method (see Section 7.2), is determining the volume of the skeleton of coarse particles and the voids between them available for the remaining SMA elements. Determining the volume occupied by the coarse aggregate skeleton consists of defining its density and testing the coarse aggregate compaction (namely, the amount of air voids remaining among the coarse grains after compacting). As we know, the amount of air voids in a compacted coarse aggregate may be determined using the following methods:

- With dry aggregates, using the dry-rodded test after AASHTO T19 as in the U.S. method, or using a gyratory compactor, Marshall hammer, or on a vibrating plate
- Using a special lubricating agent* and chosen method of compaction

The substantial difference between the dry-rodded method and other methods is the dry compaction of aggregate used in the U.S. method and the "grease" process used in the others. Explaining this issue logically, air voids determined using the dry-rodded method must be larger because of the higher resistance to the displacement of

---

* For instance, using a small amount of oil with the density at the room temperature corresponding to the density of binder at the temperature of mixing with an aggregate. So an oil with a density of ca. 0.2 Pa.s at 25°C should be used.

particles relative to each other, so they will not be arranged as closely as in the grease method. The use of grease will also result in less crushing of the aggregate during compaction. After all, when compacting SMA on a site, the presence of binder in the mixture lubricates it, making the displacement of aggregate particles easier. So it seems that the method using grease, though more problematic in practice, enables us to obtain results closer to reality. Any substance with a viscosity resembling the viscosity of binder at about 150°C may be employed as a lubricating agent or as a grease in the mixture. In the Netherlands, medical oil has been used for that purpose, with 1.5% (m/m) added to the aggregate mixture. The possibility of conducting the whole operation at room temperature, without heating up the oil and aggregate, is a notable advantage of this substance.

The coarse aggregate of an analyzed mixture (a sample of 4 kg) is compacted by being placed in 150 mm diameter specimen mold and undergoing 300 rotations in a gyratory compactor* with the external angle of rotation set on 1°. After the density of the coarse-aggregate particles (greater than 2 mm) and the air voids in the compacted aggregate are determined, the aggregate is extracted from the oil and the gradation is measured. Consequently, apart from the result of air voids in compacted coarse aggregates, some additional information is gained on the aggregates' resistance to crushing. A compactor, when set at 300 rotations, causes overcompaction of the mixture and, to some degree, destruction of grains corresponding with the laydown and compaction process and after several years of service.

Another important factor that should be taken into account while analyzing air voids among coarse grains is their crushing and wearing, which occurs at the production stage of a mixture, at its laydown, and during its later service. Crushing of the grains causes the displacement of particles, hence a decrease of air voids in the coarse aggregate skeleton (post-compaction). Coupling this with the knowledge of susceptibility of the aggregate to crushing and wearing that is gained by screening the aggregate after compacting it in the gyratory compactor, it is necessary to increase the air voids in the designed SMA to a certain degree (e.g., to 5% instead of 4%). This should guarantee that, even after long-term service under heavy loads, there will be no bleeding of mastic (fat spots) from among the grains of a skeleton. This method of reasoning has been adopted in the Netherlands, where the air void content in compacted laboratory samples for heavy duty traffic has been increased to 5% (v/v) (Jacobs and Voskuilen, 2004; Voskuilen, 2000).

The use of aggregates susceptible to crushing alters the volume relationships in the aggregate mix in the following ways:

- The volume of the coarse particle skeleton decreases (because some of the coarse particles becomes fine particles).
- The volume of fine particles, which are not involved in the coarse skeleton's performance, increases.

---

* In the Netherlands a gyratory compactor is used for such testing because the Marshall hammer is regarded as being excessively damaging to particles making the skeleton, and for that reason not fit to be used for compacting the aggregate with such a strong skeleton.

- The content of air voids in the aggregate mixture decreases.
- Consequently the quantity of air voids in the SMA drops, so the risk of overfilling with mastic increases.

Using this method, the effect of increasing air voids in the coarse skeleton has been taken into account. We should remember that air voids have been determined using the method of dry or greased compaction of coarse aggregates. In this test, only the aggregate greater than 2 mm has been used. In a real mixture, coarse grains are coated with mastic, so naturally there are particles of filler or crushed sand among the coarse particles. These particles slightly increase the content of air voids among the coarse aggregates, particularly at the first stage of an SMA's performance. In the Netherlands the effect of an added increase of air voids among coarse particles has been called *the enlarging effect*. With the passage of time, the decrease of air voids among coarse aggregates occurs as a result of post-compaction, reorientation of coarse grains, wear, and the movement of fine particles.

7.4.1.2 Fine Aggregate and Filler

According to concept of van de Ven et al. (2003), an SMA mixture probably has no real stone skeleton immediately after compaction. A real skeleton in SMA is created during service under the effects of traffic and climatic loading when sand and filler grains between the coarse aggregates (skeleton) may be crushed or moved. Accordingly, at the design stage, the content of fine aggregate and filler must be determined.

Cause-and-effect relationships between the filler and the fine aggregate (crushed sand) have not been determined in a design method. Some Dutch research into this has led to establishing the optimal relationship between those elements. Research on sand-filler mixes and the *filling* and *replacing* effects occurring between them have been elaborated on in a Dutch publication (Voskuilen, 2000). This effect is shown in Figure 7.13; its description is as follows:

- The compacted fine aggregate (crushed sand) contains a quantity of air voids.
- As filler (particles smaller than 0.063 mm) is gradually added, it fills the air voids in the sand, and the voids in the sand-filler mix get smaller. This is the *filling*-stage; the existing skeleton of the mix is made of sand (the shaded area in Figure 7.13);
- The decrease of air voids continues until the voids in the sand are completely filled (reaching the minimum possible); then only air voids in the filler remain;
- The further addition of filler with a simultaneous decrease in the amount of sand causes a gradual increase of air voids in the mix. This is the *replacement* stage; the existing skeleton of the mix is made of filler (the clear area in Figure 7.13), and grains of sand are being shoved aside by filler particles.

We have found the root of the aforementioned effect—after all, it accompanies the supplementing of fine aggregate to coarse grains in SMA—from the gradual

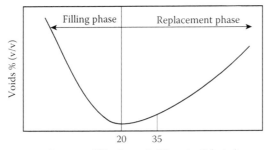

Amount of filler in sand–filler mix, % (m/m)

FIGURE 7.13   Stages of filling and replacing in a sand-filler mix. (From Voskuilen, J.L.M., Ideas for a volumetric mix design method for Stone Mastic Asphalt. Proceedings of the 6th International Conference Durable and Safe Road Pavements, Kielce [Poland], 2000. With permission.)

decrease of air voids, through the void's minimum, up to the gradual skeleton opening (as described in Chapter 6 in Section 6.2.3 on binary systems).

Looking at the form of the example in Figure 7.13, which shows the connection between the filler quantity and the content of air voids, we can observe that the rate of decrease of air voids is faster in the filling phase than its increase in the replacement phase. Thus, in the case of necessary adjustments to the content of air voids, a change in the sand fraction content will produce a stronger effect than will altering the filler content (leaving the chipping fraction unchanged) (Voskuilen, 2000).

According to Dutch research (Voskuilen, 2000), the recommended ratio of the quantity of fine aggregate (sand) to the quantity of filler amounts is 65:35 (m/m). If the filler density is about 2.700 g/cm^3 and the density of crushed sand is about 2.650 g/cm^3, this proportion may be employed without recalculation. Mass proportions should be converted into volume proportions in cases of significant deviations from these density values.

We were aware of air voids among coarse aggregate some time ago; now that we have set the sand–filler ratio, we can determine the total mastic volume in the SMA. The mastic volume is calculated according to the formula (Voskuilen, 2000)

$$V_m = \frac{m_b}{\rho_b} + \frac{m_f}{\rho_f} + \frac{m_s}{\rho_s} + \frac{m_a}{\rho_a}$$

m_b = Binder mass, % (m/m)
ρ_b = Binder density, g/cm^3
m_f = Filler mass, % (m/m)
ρ_f = Filler density, g/cm^3
m_s = Sand fraction mass, % (m/m)
ρ_s = Sand fraction density, g/cm^3
m_a = Stabilizer (drainage inhibitor) mass, % (m/m)
ρ_a = Stabilizer (drainage inhibitor) density, g/cm^3

The filling ratio stone skeleton (FRs) is used in the Dutch method to determine the theoretical degree of filling of the air voids in the coarse-aggregate skeleton with mastic (i.e., for investigating whether the design mastic volume is an optimal one). FRs is defined with the formula

$$FRs = \frac{V_m - V_s}{V_s} \cdot 100\%$$

FRs = Percentage ratio of filling the coarse-aggregate skeleton with mastic, % (v/v)
V_m = Mastic volume, % (v/v)
V_s = Air voids in the compacted coarse-aggregate skeleton, % (v/v)

Air voids in the compacted coarse-aggregate skeleton (Vs) are calculated using the formula

$$V_s = \frac{\rho_g - \rho_g^b}{\rho_g} \cdot 100\%$$

ρ_g = Density of the coarse aggregate fraction, g/cm³
ρ_g^b = Bulk density of the coarse aggregate fraction compacted in a gyratory compactor with a lubricating agent, g/cm³

The assessment of FRs ratio is as follows:

FRs < 0 implies that the air voids are not filled with enough mastic.
FRs = 0 implies that the air voids are filled with the mastic.
FRs > 0 implies that the air voids are overfilled with mastic.

For every SMA design, the FRs ratio should not exceed 0. One should remember that this is a theoretical factor and does not take into consideration the enlarging effect of the increasing air voids in the coarse-aggregate skeleton. Due to this, compacted SMA mixtures with air void contents of 4–5% (v/v) can all be marked by FRs = −4 (Jacobs and Voskuilen, 2004). It is easy to see that the content of the coarse-aggregate fraction and the size of air voids in the coarse-aggregate skeleton are dependent on the FRs level.

7.4.1.3 Binder and Stabilizer (Drainage Inhibitor)

A constant, fixed binder content, exclusively dependent on the size of the maximum particle, D, in an aggregate mix is the most unusual feature of the Dutch method. When designing SMA, the binder content for a given gradation should be taken from the regulations; for example, an SMA 0/11 for heavy-duty traffic should have a binder content of 6.5% (m/m). The quantity of binder remains unaltered; it is to be matched with a proper gradation of the aggregate mix. In other words, in the Dutch method, for a specified SMA 0/11, one has to design an aggregate mix so that it will contain 6.5% of binder with air voids at the level of 5.0% (v/v)*.

In many countries in the practice of designing asphalt mixtures it used to be said that the *optimal amount of binder* has been matched with a given aggregate mix. The Dutch method recommends the reverse. One could say that during the design the *optimal aggregate mix* had been matched with a given quantity of binder.

7.4.2   PRACTICAL GUIDELINES FOR DESIGN

7.4.2.1   SMA Constituents

Coarse aggregates of 2/6 mm are not permitted in an SMA mixture with a gradation of 0/11 mm just to guarantee a gap gradation. The sand fraction has to consist of a minimum of 50% crushed stone, whereas the content of air voids according to Rigden and the *bituminous number*[†] should form the basis for selection of the filler. As in many other countries, the amount of stabilizer is based on draindown testing.

The road binder 70/100,[‡] and in special cases modified binder, should be used in SMA. No Reclaimed Asphalt Pavement (RAP) is allowed in SMA in the Netherlands.

7.4.2.2   Designing an Asphalt Mixture

Reading the recommended fixed binder content in SMA from the standard marks the beginning of design. This quantity depends on the SMA gradation; the bigger the maximum aggregate size in a mix, the lower the intended amount of binder. It is necessary to stress again that the accepted amount of binder is a constant value; hence it is not subject to change in the design stage but is to be matched with an aggregate mix.

The amount of binder is fixed prior to establishing the filler content. The content of filler (particles less than 0.063 mm) must fall within the range of 6–10% (m/m). Limestone filler is preferred; since its content of Rigden air voids is known, its behavior in the mixture is somewhat predictable.[§]

Fixing the amount of the sand fraction is based on the indicated constant ratio of this fraction to the quantity of filler. This ratio amounts to 65:35 (m/m) if the densities are comparable with the reference density.

It has been accepted in the later stages of design that the binder content and the ratio between the filler and sand are fixed and that only the coarse aggregate content is subject to change (design). Of course, the higher the coarse aggregate content, the lower the sand-filler content. As we remember from Chapter 6, for such an assumption with a variable quantity of coarse aggregate of the same origin, the VMA will be subject to change.

---

* 5.0% of air voids is used only for heavy-duty pavements.
† So-called Van der Baan's number (i.e., the bituminous number after the EN 13179-2).
‡ Binder 70/100 with Pen@25°C from 70 to 100 dmm. Former road bitumen 80/100 after the Dutch standard NEN
§ Obviously it is about quantities of *fixed binder* and *free binder*, whose definitions and significance are discussed in Chapter 3.

Finally, the content of the coarse-aggregate fraction should amount to between 72.5 and 82.5% (m/m) according to Dutch regulations, so it is increased by 2.5% (m/m) in comparison with the most frequently used requirements (70–80% [m/m]). This is directly related to the higher requirements for the air voids content in heavy-duty pavements, at the level of 5% (v/v) (in other countries such pavements typically require 3–4% [v/v]).

7.4.2.3 Preparation of Samples

Preparation of samples for an SMA recipe is conducted according to the Marshall method, with a compaction effort of 2 × 50 blows, or in a gyratory compactor, where the number of rotations are selected in such a manner that the specimen bulk density is similar to results obtained from the Marshall method. While designing SMA with the use of the gyratory compactor, the maximum density of the mix is experimentally established.*

7.4.3 DESIGN METHOD

The applied method of SMA design in the Netherlands belongs to the group of volumetric methods. However, the assumption of a constant amount of binder in a mix makes a significant difference in comparison with other volumetric methods. When any of the final SMA parameters do not comply with the requirements, only the aggregate mix is subject to change. The consecutive stages of design using the Dutch method are discussed next.

The sequence of activities during SMA design is as follows (Jacobs and Voskuilen, 2004; Voskuilen, 2000):

1. Determine the density of materials and execute an aggregate size analysis.
2. Read a constant binder content (depending on the gradation of the mix)
3. Establish an initial design of the aggregate mix (mix 1).
    3.1 Conduct tests of air voids (V_s) in the coarse aggregate fraction (grains greater than or equal to 2 mm) with a gyratory compactor and determine the degree of crushing of the coarse aggregates (analysis of material passing through a 2 mm sieve).
    3.2 Adjust the aggregate mix by adding a crushed coarse aggregate fraction to the sand fraction.
    3.3 Calculate the volume of mastic for the initially adopted FRs ratio (e.g., −4) using the formula.

$$V_m = \left(\frac{FR_s}{100} + 1 \right) \cdot V_s$$

---

* We start with 200 rotations of a gyratory compactor. Then we determine such a number of rotations that changes the sample height by 0.2% during the last 10 rotations. Density of the mix reached at the selected number of rotations of the gyratory compactor are a reference (maximum) density for calculating the content of air voids in the SMA. (Voskuilen, J.L.M. et al., 2004.)

3.4 Calculate the volumes of the filler and sand in the mix (in the ratio of 65:35 [m/m]) for the previously calculated mastic volume, taking into account the known and constant binder content (e.g., 6.5% [m/m] for SMA 11*).

3.5 Calculate the coarse aggregate fraction's volume in the aggregate mix (the sum of volumes of all constituents should be equal to 100%).

3.6 Convert the estimated volumes of SMA components into mass units.

4. Establish the mass of coarse aggregate fraction in the initial design of an aggregate mix as x% (m/m).

5. Initiate two other variants of the aggregate mix with different contents of the coarse aggregate fraction.

5.1 Mix 2: $(x - 2.5\%)$

5.2 Mix 3: $(x + 2.5\%)$

6. Estimate the sand and filler contents (using the proportion of 65:35 [m/m]) for each of mixes 2 and 3.

7. Prepare SMA Marshall samples (2×50 blows) or gyratory compacted samples and determine the air void content in the compacted samples.

8. Determine the relationship between the coarse aggregate fraction content versus the air voids in the SMA.

9. Design this coarse aggregate content (directly or through an interpolation of results, see Figure 7.14) so that the content of air voids is in accordance with the requirements presented next.

10. Estimate the FRs ratio for the design mix.

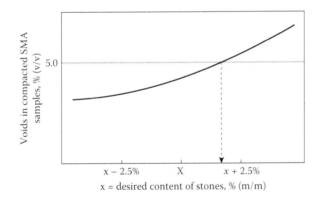

FIGURE 7.14 Example of the design coarse-aggregate fraction content based on the content of voids; $x =$ desired content of stones, % (m/m). (From Voskuilen, J.L.M., Ideas for a volumetric mix design method for Stone Mastic Asphalt. Proceedings of the 6th International Conference Durable and Safe Road Pavements, Kielce [Poland], 2000. With permission.)

---

* The bitumen content for SMA 0/8 amounts to 7.4% (m/m) for the heavy-duty traffic and 7.0% (m/m) for low-volume traffic. Both values have been calculated with 100% of the aggregate mass and the road bitumen 70/100 as a standard binder.

The requirements for air voids in compacted SMA 0/11 samples, applied in the Netherlands, are as follows:

- Type 1—for low-volume traffic
  - Marshall hammer 2 × 50: 4.0% air voids
  - Gyratory compactor: 3.0% air voids
- Type 2—for heavy traffic
  - Marshall hammer 2 × 50: 5.0% air voids
  - Gyratory compactor: 4.0% air voids

Note: the relationship between air voids obtained from the Marshall method (an impact method) and those obtained from a gyratory compactor should not be interpreted as the difference −1%, because the air void content obtained in samples from the gyratory compactor largely depends on the adopted number of rotations and the angle of rotation of the gyratory compactor. Therefore air void content is not the same in every case.

7.4.4 Application of the Method during Production Control

Subsequent to the design of the recipe and its approval, it is time for the production of an SMA mix. Production control using the Dutch method takes into account the same factors that have been used during SMA design (job mix formula [JMF]).

The properties of some elements are naturally changeable; this particularly applies to the aggregate gradation, particle shape, and resistance to crushing.

As noted earlier, designing an SMA using the Dutch method is based on optimizing the coarse-aggregate content in order to obtain the selected air void content in compacted samples of the SMA. So if the gradation of the supplied coarse aggregate changes during production of the SMA in an asphalt plant, then the SMA volumetric parameters change accordingly. Therefore the factory production control (FPC) should include examining the amount of air voids in the coarse aggregate fraction as used in the SMA design. Each new delivery of coarse aggregates should undergo such tests. The results should be compared with corresponding results of testing the materials used in the SMA design. If the difference in the air void volume is greater than 1.5% (v/v), an adjustment must be made to the design in accordance with the principles put forward in Figure 7.15.

How can we make good use of Figure 7.15? Let us fix the following values of a mix:

- The content of air voids in laboratory compacted samples amounts to 5% (v/v).
- The content of air voids in the compacted coarse aggregate fraction amounts to 36.7% (v/v).
- The content of the coarse aggregate fraction in the aggregate mix amounts to 78% (v/v).
- The FRs ratio equals 0.

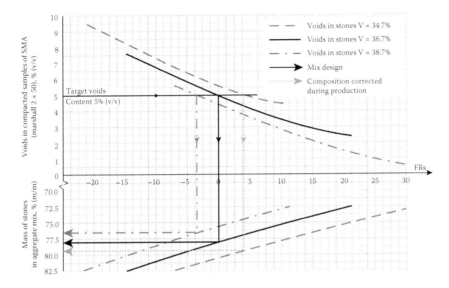

FIGURE 7.15 Graph showing the relationship between the contents of coarse aggregate fractions and air voids applied to the adjustments of mixes when producing SMAs. *Note:* FRs = Filling ratio stone skeleton. (From Voskuilen, J.L.M., Jacobs, M.M.J., and van Bochove, G.G., *Proceedings of the 3rd Eurasphalt & Eurobitume Congress Vienna 2004— Paper 326*, 1802. With permission.)

Let us assume that the properties of new deliveries of aggregates have altered slightly and are marked by a different content of air voids. To maintain the fixed volumetric relationships in the mix under these circumstances, the content of the coarse-aggregate fraction in the SMA design must be changed. Thus we test the new delivery of the coarse-aggregate fraction in a gyratory compactor,* being sure to test not only one single fraction but the entire newly composed material larger than 2 mm. Let us examine two different types of results of this testing:

- Case 1: the content of air voids in the new compacted coarse-aggregate fraction amounts to 34.7% (v/v) (i.e., 2% [v/v] less than in the design).
- Case 2: the content of air voids in the new compacted coarse-aggregate fraction amounts to 38.7% (v/v) (i.e., 2% [v/v] more than in the design).

Whether we deal with the Case 1 or Case 2, the air voids should equal 5% (v/v) of the SMA mix. So let us have a look at Figure 7.15. We can read the instructions on how the content of coarse aggregate fraction in the aggregate mix should be adjusted for the given contents (34.7% and 38.7%) of air voids in the coarse aggregate fractions and air voids in SMA (5.0%). Consequently we find:

---

* With a wetting agent.

- Case 1: the content of air voids in the compacted chipping fraction amounts to 34.7% (v/v), so the coarse aggregate content should be increased to approximately 79.5% (m/m).
- Case 2: the content of air voids in the compacted chipping fraction amounts to 38.7% (v/v), so the coarse aggregate content should be decreased to approximately 76.5% (m/m).

Adjustments to the mix make sense. If there are fewer air voids in the coarse aggregate fraction (particles are packed better in a volume unit), it is necessary to increase the coarse aggregate content ("to open" the mix) in order to retain the 5% of air voids in the SMA. If the coarse aggregate mix is more open after compaction than during design, the quantity in the mix should be reduced (it is necessary "to close" the mix). As we can see, this is the same principle that may be applied to adjust two fillers with different contents of air voids according to Rigden.

7.4.5 SUMMARY OF THE METHOD

The determination of air voids based on compacted Marshall samples has been regarded in the Netherlands as the weakest point of the Dutch method of design. According to many engineers, these samples do not reflect the true arrangement of coarse aggregate particles in real pavement. Under real-world conditions, displacement of grains and their close arrangement occur as a result of the high temperature of the pavement and post-compaction; this may be followed by the reduction of air voids among the particles that create the skeleton. Then a significant decrease of air voids in the pavement can result, right up to a complete filling-up with mastic. Subsequently, with a lack of space between the coarse aggregates, they may be shoved aside; the loss of interparticle contact and a bleeding of the mastic onto the surface of a wearing course may occur. Because of that, among other considerations, SMA for heavy traffic is designed with air voids amounting to 5% in the Netherlands, not 3–4% as in other countries.

Factory or production control adopting the principles depicted in Figure 7.15 appears to be an interesting solution; however, it necessitates the determination of relationships shown there at the design stage, which appears to be a rather time-consuming procedure. Such activities in a laboratory are reasonable provided that there is a putative guarantee of using aggregates of the same origin over the course of at least one season—that is, an unchanged SMA design will be used for a long period. Then one could afford to carry out such thorough tests. The FPC system requires a new series of tests in case of a change of an aggregate supplier.

The aforementioned method outlined in Section 7.4 describes the state of SMA design in the Netherlands in 2004. A series of research studies were conducted there in 2004–2006, leading to a change in the design procedure. Consequently the process of design was simplified in 2007. The new cycle of SMA design is as follows:

- Calculate the expected SMA void content based on volume properties.
  - Gradation of the aggregate
  - Voids in a compacted coarse aggregate skeleton
  - Binder content

- Calculate the FRs ratio.
- Calculate an actual SMA void content based on an additional corrective factor (shift factor), which denotes the relationship between the expected and actual SMA void contents.
- Assess an analyzed SMA composition based on the results of the actual air void content, adjust composition if necessary, and redo the calculations.
- Produce extra mixtures of $x + 2.5\%$ and $x - 2.5\%$ of the coarse aggregate fraction, followed by an assessment of their parameters.

7.5 OTHER METHODS OF DESIGN

7.5.1 DILATION POINT METHOD

The dilation point method, which was devised by the American National Center for Asphalt Technology (NCAT) and then adopted in Australia (Stephenson and Bullen, 2002), serves to determine the maximum content of aggregate less than 4.75 mm (i.e., the passive fraction), which still does not cause the dilation of the coarse aggregate (i.e., the active fraction). The method consists of preparing a series of samples with various contents of fine aggregate. The samples are compacted in a gyratory compactor with a constant content of binder and stabilizer. According to the rule, voids among the coarse aggregate are gradually filled with passive particles. This has an impact on the SMA's particular properties; in the Australian method the resilient modulus is examined. The point at which a skeleton is filled with passive particles is determined through an analysis of the height of samples during compaction in the gyratory compactor. Tests of different contents of fine aggregate at various screenings through 4.75, 2.36, 1.18, and 0.6 mm sieves have been conducted.

7.5.2 IRISH METHOD

In the publication by Brennan et al. (2000) is proposed another method of SMA design developed in Ireland. This method allows for a required discontinuous gradation of an asphalt mix between 0.6 mm and 5.0 mm sieves (for SMA 0/14) to create a strong skeleton of coarse grains. The determination of the properties of the coarse aggregate fraction (greater than 2 mm), screened out of the aggregate mix, has been recognized as the key issue, as in other methods. The density of the compacted coarse aggregate fraction is determined by vibration; a sample is tested under low pressure in the mold used during the California Bearing Ratio (CBR) test by placing it on a vibration table commonly used for compacting samples of ready-mix concrete. The sample is compacted to the refusal density with the use of variable values of amplitude until the moment particles start crushing (the amount of fine particles smaller than 2 mm is about 1%).

Having already obtained information about the density of the coarse-aggregate fraction and the target content of air voids in the SMA samples, one can use the formulas detailed by Brennan et al. to calculate the necessary content of the coarse-aggregate fraction for the mix to fulfill all determined requirements (e.g., void content).

7.5.3 Bailey's Method

The Bailey method was created in the United States in the early 1980s by Robert Bailey of the Illinois Department of Transportation. It enables the selection of an aggregate gradation that guarantees the best interlocking of aggregate particles, a suitable amount of VMA, and proper voids in the final asphalt mix. It was primarily intended for designing continuously graded mixtures with high deformation resistance, but it may also be applied to designing SMA gradation.

The method for gradation selection is based on the principle of the packing characteristics of aggregates and finally allows designing mixtures with expected aggregate interlocking. The complete method is used only for aggregate gradation design, not for full SMA (with binder) recipe design.

Those who are interested in Bailey's method can refer to the publication by Vavrik et al. (2002), which contains a very detailed description of the consecutive stages of the procedure, calculations, and some examples of design.

7.5.4 Method of Successive Iterations

This method, also called the method of iteration, is known chiefly by those who have practiced ready-mix concrete design in a laboratory. It is a laborious procedure because numerous tests need to be conducted (so now it is rarely used). It involves selection of consecutive aggregate fractions so that smaller particles can fit in among bigger ones with no increase in the mix volume (i.e., so that no larger particles are shoved aside by smaller ones).

We start by fixing the proportion of the coarsest aggregate (N_1) with a finer fraction (N_2). Having found such proportions for which we have a minimum of air voids with an unchanged volume of the mix, we mark the acquired mixture as aggregate $N_{1,2}$. Then we start establishing the proportion of the aggregate $N_{1,2}$ with the finer aggregate N_3, and so on. The process involves making consecutive batches of aggregate and determining air voids for each mix. Its comprehensive description has been explained in the publication by Śliwiński (1999).

Despite the completely different primary purpose of this method, it seems that nothing stands in the way of making use of its principles for determining a suitable gradation of an SMA aggregate mix.

8 Analyses and Laboratory Tests

The components of stone matrix asphalt (SMA) and methods of SMA design have been discussed in the previous chapters of this book. In addition to the basic mix design processes, however, there are additional laboratory tests and procedures that are recommended to supplement the mix design work. Some unique tests focus on special properties of SMA mixes that may be encountered from time to time. Therefore it is necessary to undertake a short review of some selected tests. Some guidance is also offered on the interpretation of test results to avoid misunderstandings caused by inaccurately defined properties or incorrect test parameters. Such cases are also described in this chapter.

8.1 PREPARING SAMPLES IN A LABORATORY

At present, two methods of preparing samples in a laboratory are commonly applied worldwide: the use of a Marshall hammer or a gyratory compactor. These methods and the problems related to them will be discussed here. Comparisons of other laboratory methods of compacting samples may be found in several publications (Gourdon et al., 2000; Hunter et al., 2004; Renken, 2000).

8.1.1 PREPARING SAMPLES WITH THE USE OF THE MARSHALL HAMMER

The procedure of preparing samples using the Marshall method is common. It is currently available in EN 12697-30 and ASTM D 6926-04.

The biggest drawback of this method is the incompatibility of laboratory conditions and the realism of a construction site. This type of laboratory compaction consists of tamping a mix down with strokes of a predetermined compaction effort defined by the number of impacts on the face of a cylindrical sample.* The Marshall method was created several decades ago for designing an optimal content of binder in fine- and medium-graded mixes with continuous grading. Despite progress in technology, the impact method of compaction is still being widely used.

Besides the gap between the lab and the construction site, the most important issue is the energy used for compacting SMA mixtures. Let us have a look at the compaction effort of 2×50 (50 strokes on each face of a cylindrical sample) and another one of 2×75. Greater compactive efforts—namely, 2×75—are typically applied when designing asphalt concrete mixes for courses under heavy traffic. Such

---

* In this book that energy is usually referred to in $2 \times A$ conventional notation, where A denotes the number of impacts on one side of a sample (e.g., 2×50 or 2×75).

considerable compaction efforts may also be used for mixes of continuous grading. Gap-graded mixes with a strong coarse aggregate skeleton, like an SMA, may experience the following two subsequent stages of compaction:

- Compaction of the skeleton until stone-to-stone contact is achieved
- Additional compaction, which may cause the crushing of weaker grains (overcompaction)

The proper compaction of an SMA mix is achieved at the moment when the stone-to-stone contact is reached, which corresponds to a certain amount of compactive effort. That is why in most countries (e.g., Germany, the United States) the compactive effort of 2 × 50 has been determined as adequate, regardless of the traffic assignment. Though it happens rarely, in some countries the same rules are used for the specification of SMA as for asphalt concrete, specifically 2 × 50 for low- and medium-traffic loading and 2 × 75 for heavy traffic.

In some countries, additional tests of SMA sensitivity to overcompacting are required. In the Czech method (see Chapter 7), typical compaction (2 × 50) is used for design, and an additional test of 2 × 100 strokes is used to indicate the resistance of the aggregate mix to overcompaction. It is obvious that after such overcompacting some of the aggregates will be crushed. The requirement for the air void content (more than 2.5% with 2 × 100) in such overcompacted samples could help ensure that even in cases of overcompacting during rolling or when weak aggregates are used, the SMA layer will still have enough void space between the aggregate grains.

Research conducted in many countries (Boratyński and Krzemiński, 2005; Brown and Haddock, 1997; Perez et al., 2004) into the compaction of SMA has demonstrated clear evidence that an excessive compaction effort put into Marshall samples leads to aggregate crushing and adverse volume changes in the mix.

8.1.2 Preparing Samples with the Use of a Gyratory Compactor

The use of a gyratory compactor is another popular method worldwide for preparing laboratory samples. This piece of equipment is not new; this method of compacting samples was developed in the late 1930s and early 1940s.* Making samples with the use of the gyratory compactor is described in EN 12697-31, ASTM D4013-09, and ASTM D3387-83(2003).

This method of compacting samples consists of kneading a mix with a rotational force. The crucial features of the gyratory compactor are

- Angle of rotation
- Vertical pressure
- Number of gyratory rotations

---

* The first experiments with this type of apparatus were performed from 1939 to 1946 by the Texas Highway Department in the United States. More about the history behind the gyrator compactor can be found in Harman et al. (2001).

- Initial ($N_{initial}$)
- Design (N_{design})
- Maximum (N_{max})

The air void content after an initial number of rotations (usually 9 or 10) are a measurement of the compactability of a mix.

There are several types of such instruments, of a dozen or so makes, with more than 3000 examples of them in laboratories. The following are distinct types of these devices:

- Superpave Gyratory Compactor (SGC) is used in the United States for designing mixes using the Superpave method (AASHTO T312 standard) and for designing SMA. Settings according to AASHTO T312 are as follows:
  - Internal angle of rotation: 1.16° ± 0.02°
  - Vertical pressure: 600 ± 18 kPa
  - Rotational speed: 30 ± 0.5 rev/min
  - Diameter of sample: 150 mm
- Gyratory Testing Machine (GTM) is the press that was designed and built by the U.S. Army Corps of Engineers.
- Presse de Compactage à cissaillement Giratoire (PCG) is the press that was designed and built by the LCPC* in France; its parameters have been adapted to the guidelines of various methods, inter alia Superpave method.

In both the Marshall and gyratory methods of compaction, the important point is fixing suitable design parameters precisely. In the recent past, the number of rotations of a gyratory compactor have been established as a parameter corresponding with the number of strokes of the Marshall hammer (to determine density and air void content of a given mix). According to the results of research presented in Brown and Cooley (1999), the number of rotations ($N_{design} = 70$ or 100) used for an SMA design with an SGC are equivalent of the Marshall hammer compactive effort 2 × 50. The number of rotations chosen depends on the resistance to crushing (Los Angeles [LA] index) of the coarse aggregate used; 70 revolutions of an SGC should be assumed when the LA index amounts to 30–45% and 100 revolutions for the LA index less than 30%. In other documents, like NAPA SMA Guidelines QIS 122, the $N_{design} = 75$ and 100 rotations are used for SMA design with an SGC. The current AASHTO R46 standard corresponds with the NAPA publication (75 and 100 gyrations according to the LA index), values that are used in the United States.

The Australian guidelines NAS AAPA 2004 use a different compactive effort. During SMA design for low- to medium-traffic loading and for heavy- to very heavy-traffic loading, the preferred values are 80 and 120 rotations of the compactor, respectively.

---

* Laboratoire Central des Ponts et Chaussées, or Central Laboratory for Roads and Bridges, France

8.1.3 Visual Assessment of Laboratory Samples

A clause in one of the Polish documents on SMA design (ZW-SMA-2001) reads as follows:

> After preparing Marshall samples, an additional assessment can be done.... A visual assessment of samples should be carried out: coarse aggregates should be noticeable on the surface of the sample, while voids between them should only be partially filled with mastic...

That sounds simple and clear, so let us have a look at a few examples of samples. Figure 8.1 shows an image of a proper SMA sample. There is no excess of mastic, coarse aggregates are visible, and the voids between them are "only partially filled with mastic." This is how a properly designed and compacted SMA specimen should look.

Figure 8.2 depicts a remarkably different image of SMA samples, though maybe some readers can hardly believe it is still SMA. It really looks like mastic asphalt. There are few, if any, coarse aggregates standing out; voids only partially filled (with mastic) are not easy to find either. Generally, that SMA design could be immediately disqualified, but before rejecting the recipe of Figure 8.2, it is definitely worth giving more thought to the practical reasons of that unsuitable appearance. Mastic is squeezed out, which means there was too much of it in comparison with the free space in the aggregate mix or maybe there were too few air voids in the aggregate mix. Consequently a mistake at the stage of mastic design was made (in which case the mix should be designed once again) or SMA samples were improperly compacted (with too great a compaction effort). Maybe the aggregate was too weak and was subsequently crushed during compaction in the mold.

Visual assessment plays only a small role, because human senses may be deceived. After all, we can imagine an SMA sample with a low binder content but compacted with a mighty effort. At that time the sample may look fine—we are under the illusion that the mastic–binder content is sufficient. But in the field, a comparable compactive effort cannot be applied. As a result, the layer will turn out to be open and

(a)

(b)

FIGURE 8.1 The appearance of a Marshall sample of an SMA mix after compaction and removal from the mold: (a) frontal plane and (b) lateral plane. (Photos courtesy of Alicja Głowacka.)

(a)   (b)

FIGURE 8.2   The appearance of Marshall samples of an SMA mix after compaction and removal from the molds: (a) and (b) are frontal plane views of SMA samples with an excess of mastic. (Photos courtesy by Krzysztof Błażejowski.)

porous. So, despite the good appearance of the lab sample, the mix will not perform successfully in the field. As they say, sometimes looks can be deceiving. On the other hand, the method of visual assessment is sometimes useful, as long as the selected parameters of sample preparation are correct.

8.1.4   Preparing Samples with a Granulated Stabilizer

An unexpectedly high draindown may be obtained during laboratory tests in which components, including a granulated stabilizer, are mixed with a small mixer (or manually). This may be caused by the way the granulated stabilizer was prepared before mixing it with aggregates. As we know from Chapter 4, the granules contain a small amount of binder or wax. This coating makes them less sensitive to moisture and makes dosing easier. However, the coating also requires enough shear force and a high enough temperature to release the fibers during mixing.

To avoid trouble with dispersing the granulated fibers, the container with a weighed-out amount of granulated stabilizer should be put in the oven before mixing and warmed up to a temperature at which the binder or wax in the stabilizer clearly softens. This will make distributing the fibers through the mix easier. In an unheated granulated stabilizer, some granules may remain intact; therefore the stabilizer will not work effectively. The effect will be self-evident—an incorrect (high) result of draindown testing.

Be careful when testing stabilizers with which you are unfamiliar. Learn about appropriate mixing temperatures and mixing details, such as whether the stabilizers should be added to the dry aggregate or to the mixture with binder.

8.2   DRAINDOWN TESTING

The draindown effect is the process of the separation of liquid binder or mastic from an SMA mixture that occurs at a high temperature when the binder is still molten.

Both the binder and the mastic can separate. It is common knowledge that SMA mixes are marked by an intentional excess of binder, and the draindown effect is caused by the impossibility of maintaining such an excess of binder on grains of aggregate. That problem is most frequently solved by adding a stabilizer (or drainage inhibitor) to a mix. Its task is to absorb any excess of binder.

Excessive draindown may be caused by several factors (as described in Chapter 4). Binder or mastic draindown from a mix brings about many problems; most of them are described in Chapter 11.

Research on an SMA mix's susceptibility to binder or mastic segregation has been conducted in virtually all the countries using this type of mix. A short review of the procedures of draindown testing and the regulations for the following methods will be discussed later:

- Using Schellenberg's method (the original German method)
- Using the European standard EN 12697-18
- Using the U.S. standard AASHTO T 305-97

8.2.1   Original Schellenberg's Method

A method of mastic draindown testing was published in 1986 by Kurt Schellenberg and Wolfgang von der Weppen (Schellenberg and Weppen, 1986). Their original method, which consisted of warming up a sample of SMA mix placed in a glass beaker, is summarized in Table 8.1.

TABLE 8.1

Parameters of Draindown Testing according to Schellenberg's Method

Number of samples	1
Test temperature	170°C ± 1°C
Sample weight	1000–1100 g
Test time duration	60 ± 1 minutes
Test procedure	1. Warm-up an empty beaker in an oven at test temperature, take it out and weigh it, and put it into the oven.
	2. Mix SMA components.
	3. Remove the beaker from the oven, quickly put a prepared mix into the beaker, and weigh them altogether.
	4. Place the beaker with the mix in the oven for 60 ± 1 minutes.
	5. Remove the beaker with the mix from the oven and empty the mix by tilting the beaker upside down.
	6. Weigh the cooled beaker with remaining mastic to an accuracy of 0.1 g.
	7. Calculate the draindown as a ratio of the mass remaining in the beaker to the original SMA mass and express result as a mass percentage.

Source: Based on Schellenberg K. and von der Weppen W., Verfahren zur Bestimmung der Homogenitäts-Stabilität von Splittmastixasphalt. Bitumen, 1, 1986.

8.2.2   Method after AASHTO T 305-97

The U.S. method of mastic draindown testing has been described in the standard AASHTO T 305-97. It is used for porous asphalt mixes (also called open-graded friction course [OGFC]) and SMA mixes. Test parameters are shown in Table 8.2.

Samples of the mix are placed in wire baskets (Figure 8.3). For SMA mixes equal to and larger than 9.5 mm maximum aggregate size, the basket should have 6.3 mm holes in the mesh, and for 0/4.75 mm SMA mixes, the holes should be 2.36 mm.

8.2.3   Methods after EN 12697-18

The two methods of draindown testing that are given in the European standard EN 12697-18 are the method with a basket and Schellenberg's method. The method with a basket after EN 12697-18 (Part 1) is mainly used for draindown testing of porous asphalt. In principle, it is possible to determine only binder draindown but not mastic

---

TABLE 8.2
Draindown Test Parameters according to AASHTO T 305-97

Number of samples	Four total; test two samples of a mix at each of the two test temperatures.
Test temperature	Samples are to be tested at two temperatures:
	1. The expected production temperature in an asphalt plant (two samples)
	2. A temperature higher by 15°C than the expected production temperature (two samples)
Sample weight	1200 ± 200 g
Test time duration	60 ± 5 minutes (or 70 ± 5 minutes in case of oven cooling)
Test procedure	1. Weigh the tray to catch flowing mastic to an accuracy of 0.1 g.
	2. Mix components of an SMA mix at a fixed temperature.
	3. Put the prepared mix into a weighed wire basket; do not pack the mix into the basket, and do not postcompact it either.
	4. Measure the basket mass with an accuracy of 0.1 g.
	5. Check the temperature of the mix; it should not drop by more than 25°C below the desired test temperature. If it does cool too much, the mix should be kept in the oven for 10 minutes longer, (i.e., up to 70 minutes).
	6. Place the basket with the mix on the tray, and then put it into the oven for 60 ± 5 minutes.
	7. Remove the tray with the basket from the oven, and weigh the basket with the mix or the tray itself with an accuracy of 0.1 g.
	8. Determine draindown as a percentage of the mastic mass remaining on the tray in relation to the total mass of the mix before testing.
Remarks	• When stirring the mix components, pay attention to the proper sequence of their dosages, particularly fibers, polymers, and so on.
	• The temperature of aggregate in the oven (before mixing) cannot exceed the desired production temperature of a mix by more than 28°C.
	• The final result is the arithmetic average of the two samples at each test temperature.

FIGURE 8.3   The wire basket for draindown testing according to AASHTO T 305-97. (Photo courtesy of Karol Kowalski and Adam Rudy, Purdue University.)

because the basket used here has small perforated holes. Moreover, these holes may be blocked during the testing of mixes containing larger quantities of mastic and fiber stabilizers. This is the reason why that method has limited application for SMA draindown testing.

Schellenberg's method according to EN 12697-18 (Part 2) has been applied in draindown testing of porous asphalt containing fibers and other asphalt mixes like SMA. The essential information on that method is shown in Table 8.3.

8.2.4   SUMMARY AND COMMENTS ON DRAINDOWN TESTING

The methods of draindown testing described in this chapter differ in details. Table 8.4 shows the most important differences among them. Table 8.5 depicts commonly adopted assessment criteria of draindown testing results.

The following remarks deserve mention:

- The fixed temperature of draindown testing adopted in the original Schellenberg's method, 170°C, represents an average SMA production temperature for a typical mix with binder having a Pen@25°C = 50–70 dmm. However, it is necessary to note that higher temperatures have been used

TABLE 8.3
Drain-Off Test Parameters for Schellenberg's Method according to EN 12697-18

Number of samples	Test three samples of the same mix with the same binder content.
Test temperature	Test temperatures depend on the binder type: 1. For road binder—at the production temperature of a mix defined according to EN 12697-35 and raised by 25°C 2. For modified binder—at the production temperature of a mix defined by the binder supplier and raised by 15°C
Sample weight	• Mass of an aggregate mix sample—1000 g for a mix with the density of 2.65–2.75 g/cm³ • If the density of an aggregate mix is different from the given reference density, a sample mass should be calculated to obtain the same test material volume
Test time duration	60 ± 1 minutes
Test procedure	1. Prepare three batches of aggregate (batches 1, 2, 3) and place batch in a metal container. 2. Put the beakers in the oven at the test temperature for 15 minutes minimum; then remove them, weigh them with an accuracy of 0.1 g, and return them to the oven. 3. Mix 1 kg of a bituminous mixture at the fixed temperature according to EN 12697-35. 4. Remove the beaker for batch 1 from the oven, quickly put the prepared mix in the beaker, weigh the beaker with the mix with an accuracy of 0.1 g, write down the time and the beaker number, and return the beaker to the oven (it should not be left outside the oven for longer than 60 seconds) 5. Prepare the two remaining batches of the mix in the same way and put them in the beakers. 6. Keep each beaker with the mix in the oven for 60 ± 1 minutes. 7. Remove the first beaker with the mix, measure its temperature, and put the mix aside. 8. Remove the remaining two beakers with the mix from the oven and empty them out by tilting them upside down and holding them in that position for 10 ±1 seconds. 9. After cooling beakers No. 2 and No. 3, weigh them together with the remaining binder with an accuracy of 0.1 g. 10. If more than 0.5% of the initial mass of the mix remains on the beaker walls (including aggregate grains and mastic), the material remaining in the beaker should be washed with the solvent and passed through a 1-mm sieve; next, the material remaining on the sieve should be dried and weighed with an accuracy of 0.1 g. 11. Determine draindown as the percentage of the binder mass remaining in the beaker compared with the mass of the mix. 12. Calculate the material draindown, D, and when appropriate, the material remaining on the 1-mm sieve, R:

$$D = 100 \times \frac{\left(W_3 - W_1 - W_4\right)}{\left(W_2 - W_1\right)}$$

(Continued)

TABLE 8.3 (CONTINUED)
Drain-Off Test Parameters for Schellenberg's Method according to EN
12697-18

$$R = 100 \times \frac{W_4}{\left(W_2 - W_1\right)}$$

where

D = Material draindown (% m/m)
R = Material remaining on the sieve 1.0 mm (% m/m)
W_1 = Mass of the empty beaker (g)
W_2 = Mass of the beaker with the mix (g)
W_3 = Mass of the empty beaker together with remaining mastic (g)
W_4 = Mass of the dry material remaining on the 1.0-mm sieve (g)

13. The average result of two measurements should be given with an accuracy of 0.1%.
14. Results for D and R (if applicable) should be reported.

Remarks
- None of the three beakers containing the mix may be kept in the oven for longer than 60 ± 1 minutes.
- While mixing components of the mix, pay attention to the proper sequence of their dosages, particularly fibers, polymers, and so on.
- In the case of modified binder, a lot of mastic can stick to the walls of the beaker and remain there when emptied out (due to increased tackiness of a mix). In such a case, retesting should be conducted at a temperature 5°C higher. If the new result is lower than the previous one, it should recorded in a report.
- If the difference between the test results for two samples of the same mix with the same binder content exceeds 0.5%, a new pair of samples should be tested.

for the production of SMA with polymer modified binder. Moreover, in cool or cold months asphalt mixes are usually produced at temperatures that are slightly higher than 170°C.

- It should be emphasized that the AASHTO procedure does not provide for a fixed binder draindown test temperature but makes it conditional on the expected SMA production temperature at an asphalt plant. Such an approach has its advantages. Above all, it ensures that draindown will not occur at the real SMA production temperature, thus the risk is easily estimated. Additionally, draindown testing at a temperature higher than the production temperature (by 15°C) makes that certainty even stronger since sudden temperature fluctuations happen sometimes, particularly at the beginning of production. Similarly, the European standard EN 12697-18 has defined the draindown testing temperature as 15–25°C higher than the planned mix production temperature.

- Raising the draindown test temperature may bring about an increase of stabilizer content in a mix, but to a larger degree it will reduce the risk of fat spots during laydown.

- When using granulated fibers and making mix samples without a mechanical mixer (i.e., by mixing them manually in a pot), remember to heat the

TABLE 8.4
A List of Differences among Methods of Draindown Testing

Method	Test Temperature (°C)	Test Time Duration (min)	Container
Original Schellenberg's method	170 ± 1	60 ± 1	Glass beaker
AASHTO T 305-97	Depends on the SMA production temperature and type of binder	60 ± 5 or 70 ± 5	Wire basket
EN 12697-18 (Part 2)	Depends on the type of binder	60 ± 1	Glass beaker

TABLE 8.5
Assessment Criteria of Draindown Testing Results

Drain-off Testing Result, % (m/m)	Assessment
>0.3	Risk of binder draindown
0.2–0.3	Acceptable value
<0.2	Recommended value

granulated stabilizer in an oven just prior to mixing it with aggregate (see Section 8.1.4).

- Two more issues regarding draindown testing with the use of a glass beaker are worth raising.
  - Before putting hot SMA in a glass beaker, the beaker should be heated in an oven to the test temperature, otherwise the hot mastic of SMA will easily stick to the cool walls of the beaker, falsifying the test results. Many people claim that testing with a beaker is a matter-of-fact measurement of mastic-to-glass adhesion.
  - Let us imagine pouring a mix from the glass beaker onto a tray after warming the mix in an oven for 1 hour. We have weighed all the material remaining in the beaker (i.e., the mass of the beaker with the SMA residue). The question is, is all the remaining material in the beaker the real 'potential' SMA draindown? After all, the rest of the mastic, sand, and grit grains stuck to the walls remain in the beaker. Some material from any asphalt mix would cling to the beaker even if there were no draindown. So, the residue in the beaker should not be considered completely equivalent to drained off mastic. Logic demands, however, that we recognize the material remaining on the bottom of the beaker as draindown. In light of these concerns, the method of measuring material outside a container— namely, the AASHTO method with a wire basket—deserves consideration.

8.3 FILLER TESTS

8.3.1 TESTS OF SPECIFIC SURFACE WITH THE USE OF BLAINE'S METHOD

Blaine's method is chiefly used when testing cement (grinding gradation control). It consists of the measurement of time necessary for air to flow through a compressed layer of tested material of a given size and porosity. At standard conditions, the specific surface is directly proportional to $\sqrt{t}$ (t = time of air flow). A master sample with a known specific surface is required to calibrate testing. The test can be carried out according to EN 196-6.*

8.3.2 DETERMINATION OF COMPACTED FILLER AIR VOIDS AFTER RIGDEN'S AND RIGDEN–ANDERSON'S METHODS

Rigden's and Rigden–Anderson's methods apply to any fine material used as a filler in hot mixes (e.g., bag-house fines and added filler). Filler air voids make up an air volume occurring among grains of filler compacted with a special apparatus by a standardized method. Test methods according to the EN standard (Rigden) and the U.S. procedure (Rigden–Anderson [Anderson, 1987]) differ markedly, which makes the direct comparison of results impossible. The only feature they have in common is their principle—dry compaction of filler.

8.3.2.1 Rigden's Method after EN-1097-4

The EN method provides for compaction of a dry sample of filler by 100 strokes of a dead weight every second. The mass of sample is 10 g, and the mass of the dead weight is 350 g.

The volume of air voids is estimated, taking into account the mass of the compacted sample, its volume, and the filler density. European countries have used the EN 13043 standard for aggregates for asphalt mixes. The most frequently adopted requirement for a filler tested according to EN 1097-4 is the category $V_{28/45}$, which means the content of air voids should be within 28–45% (v/v).

8.3.2.2 Ridgen's Method Modified by Anderson

The U.S. method (here called Rigden–Anderson) described in Anderson (1987) stipulates 25 strokes of the 100 g dead weight. Results of the measurements form the basis for calculating the volume of air voids in a dry compacted filler. Only 1.0–1.3 g of filler is needed to conduct the testing. After determining the content of air voids in the compacted filler, the calculation of free and fixed binder may be performed (by mass and by volume). The concept of free and fixed binder is presented in Chapter 3.

8.3.2.3 Comparison of Methods

Rigden's method (the European procedure) and Ridgen's as modified by Anderson (the U.S. procedure) produce different contents of air voids in the same compacted

---

* EN 196-6. Methods of testing cement—Determination of fineness.

filler. This is caused by a lower compacting effort in the U.S. method (25 strokes) than in the European one (100 strokes). Much research, particularly in the United States, has been carried out with the use of Ridgen–Anderson's method. The results would be very valuable for Europeans, but test conditions differ to such an extent that a simple comparison of results is rather impossible.

8.3.3 TESTS OF FILLER STIFFENING PROPERTIES

Filler stiffening properties may be tested with the following methods:

* Methods of increasing of softening point—ring and ball (R&B) method, according to EN 13179-1 (delta ring and ball), and similar methods
* Rigden's method, according to EN 1097-4, and Rigden–Anderson's method
* The method of increasing mortar viscosity

The two European tests cited in EN 13043 (EN 13179-1 and EN 1097-4) are carried out for an added filler and the 0/0.125 mm fraction sieved out of the fine aggregate (or an aggregate of continuous grading with D less than or equal to 8 mm) that contains more than 10% dust. Let us dedicate some time to discussing those tests since understanding them will help determine the expected values of a good filler.

8.3.3.1 Method of a Softening Point Difference

8.3.3.1.1 Method EN 13179-1 (Delta Ring and Ball)

What is delta ring and ball (ΔR&B)? According to EN 13179-1, it is an increase in the R&B softening point of a binder-filler mixture consisting of 37.5 parts of filler and 62.5 parts of binder by volume, related to the R&B softening point of the pure binder used for testing. The part of the filler passing through the 0.125 mm sieve and, according to EN 1259, road binder type 70/100 are used for testing. Measurements are taken according to EN 1427. The final result is denoted ΔR&B (or in simplified terms, delta).

European countries, which created their requirements for fillers according to the common EN 13043 standard, mostly require a class $\Delta_{R\&B}8/25$, which signifies a stiffening power between 8°C and 25°C.

8.3.3.1.2 Other Methods

Various methods of testing for an increase in the softening point are used in many countries. For example, in Germany, two methods are applied: the R&B method and Wilhelmi's method. According to an U.S. review of fillers (Harris and Stuart, 1995), in Germany an acceptable range of ΔR&B of 10–20°C has been adopted for the R&B method, with components selected at the filler-binder content ratio (F:B) equal to 65:35, % (v/v). Mortars with ΔR&B greater than 20°C are too stiff and are not accepted. Similarly, mortars with ΔR&B less than 10°C are not accepted due to their excessive plasticity.

Another interesting test applied in Germany is the determination of a stiffening factor (in Germany *Stabilisierungindex*) (Schellenberger, 2002). It is an F:B for which the mortar ΔR&B increase is equal to + 20.0°C. It is necessary to make a series of filler-binder mixtures with different F:B ratios (e.g., 1:1, 1.5:1, and 2:1) and then determine a ΔR&B increase compared with the pure binder R&B. As a result, we obtain a graph showing the increase versus the F:B ratio; this can be interpreted as the relationship of the stiffening power of a given filler. When studying the results of stiffening factor tests, it is taken as a general rule in Germany that the results should be higher than approximately 1.9. The lower the stiffening factor, the stronger the stiffening impact of that filler on an asphalt mix.

Basically, the results of the tests lead to the conclusion that research on the increase in a softening point does not always reveal all the negative properties of a filler (e.g., swelling). Nevertheless, their merits lie in the ease with which the softening point can be measured through the R&B method.

8.3.3.2 Method for Testing the Increase in Mortar Viscosity

Some references to a method consisting of the testing of mortar viscosity and comparing it with the pure binder viscosity may be found in the literature. By and large, such a comparison would be a stiffening factor. However, as it has been pointed out in Anderson's work (Anderson, 1987), not only do the filler properties affect that factor but the properties of the binder used for testing do as well. Certainly the reliability of that method is controversial.

8.3.4 Other Factors and Filler Tests

8.3.4.1 German Filler Test

An interesting and simple method of testing fillers is that discussed in the study by Kandhal et al. (1998). Called the German Filler Test,* it consists of determining the amount of filler required to absorb 15 g of hydraulic oil and is carried out as follows:

1. Put 15 g of hydraulic oil into a small melting pot, add 45 g of filler, and mix them together.
2. Shape the mixture into a ball.
3. If shaping the ball is successful (i.e., it does not break down) put it into the pot again and add another 5 g of the filler.
4. Having mixed both components, shape the ball one more time and inspect its cohesion.
5. Repeat with another 5 g filler batch until the ball breaks down (lost cohesion).

It is then assumed that the 15 g of hydraulic oil have been completely absorbed by the filler air voids. The test also indicates there is a lack of free oil (similar to free binder) in the mixture that might bond the mortar together. In that case, the result shows the quantity of filler (in grams) required to achieve that condition. The results of this test show a very good correlation to results from the modified Rigden test (Kandhal et al., 1998).

---

* The procedure was developed by the Koch Materials Company.

9 The Production of SMA

Having designed and checked the SMA mixture, the time has come to produce it according to the job mix formula (JMF). In this chapter we shall deal with

- Requirements for the organization of an asphalt plant
- Assumptions and control over the SMA production process
- Production of the SMA mixture in a batch plant or in a drum-mix plant
- Storage of manufactured SMA in a silo

9.1 REQUIREMENTS FOR THE ORGANIZATION OF AN ASPHALT MIXING PLANT

The organizational requirements of an asphalt mixing plant and its surroundings are much the same for the production of SMA as for other asphalt mixtures. Some special issues may arise, however, when dealing with SMA, such as the following:

- Storage of aggregates
  - Stockpiled aggregate should not be mixed with underlying-soil material.
  - Covered aggregate stockpiles may be desirable, especially for the fine aggregate stockpiles; a lower aggregate moisture content improves the plant's output; aggregates may also be stored in silos, after preliminary drying, but this is still rarely done (see Figure 9.1).
- Storage of stabilizers—covered storerooms may be used; dry storage is especially important when storing loose stabilizers (nongranulated).

Two types of asphalt plants may be singled out with regard to the manner of mixing components—batch plants and drum-mix plants. Batch plants are the most popular in Europe, whereas drum-mix plants may be seen elsewhere in the world. Drum-mix plants can be adapted for SMA production, however, they require some special solutions for batching stabilizers.

The output of a particular asphalt plant should be adjusted to the intended placement efficiency (e.g., the width and thickness of a course, the distance from the work site, and the number of trucks for transportation) in order to organize the SMA laydown so that the stops of the paver are kept to a minimum. Keeping the paver moving forward steadily helps improve the smoothness of the final pavement and limits differential cooling of the mat.

(a)

(b)

FIGURE 9.1   An asphalt plant in Slovenia with all aggregates stored in silos: (a) a view of a silo, and (b) the aggregate delivery chute into an underground chamber. (Photo courtesy of Krzysztof Błażejowski.)

It is easier to produce the SMA mixture in conformity with a job mix formula if the sieves in the screen deck of a batch asphalt plant are properly selected, making control of the mineral mix easier. Improperly selected sieves may results in too wide a hot-bin size range (e.g., 2/10 mm for SMA 0/12.5), which may cause problems with adequately controlling the mix production in accordance with the mix design.

9.2 PRACTICAL CONSIDERATIONS OF THE SMA PRODUCTION PROCESS

By and large, the production of an SMA in contemporary asphalt plants does not present particular problems. The following are a few general tips about the production of SMA:

- SMA requires some production consistency with no breaks, stoppages, or similar "jerking" of the production process. Any alterations to the type of mixture being produced require adjustments of the batching device controls, the weight of mixture constituents for the mixer, and so on. Potentially more troublesome, any stoppages necessitate restarting the machine and beginning the production again.
- The moisture content of the aggregate leaving the dryer should not be higher than 0.5%, optimally less than 0.2% (USACE Handbook, 2000).
- When initial batching (cold feeders) limits the machine's output, an additional batching device should be considered; bear in mind that coarse aggregates constitute more than 70% (m/m) of the mixture and may require more than one bin to feed that large a quantity.
- Due to the small amount of sand in an SMA, coarse aggregates passing through the dryer's drum are exposed to more intense heating; therefore it is important to make sure that the asphalt mixture is not overheated.

9.3 PRODUCTION PROCESS

Suppose that we have already developed the job mix formula and that well-performing batching devices, a screen deck, balances, and so on are at our disposal. Then we are ready to start production. A common occurrence with starting up an asphalt plant is the instability of the mixture temperature during the first production period. Therefore one should take into account that some batches will be underheated, while others will be slightly overheated. Such waste material should be rejected.

9.3.1 SMA PRODUCTION TEMPERATURE

Two components of an SMA mixture must be heated—the aggregate and the binder. This heating is aimed at (1) eliminating moisture from the aggregate to a level that enables the proper coating of the aggregate grains and (2) maintaining the appropriate temperature of the mixture delivered to the laydown site, which allows for its proper placement and compaction. The coating temperature is directly related to the viscosity of the chosen binder. The harder the binder or the more highly modified the binder is, the higher the production temperature must be. That is why the SMA production temperature is most often specified as a function of the type of binder.

The SMA production temperature has been diversely defined in different countries in the following manner:

- Variant 1—the types of binder and the recommended production temperature range are provided in one specification. This is the simplest way to reveal data and ensure substantially uniform conditions for SMA production; EN standard 13108-5 (which applies only to unmodified binder) is a good example of such a document.

- Variant 2—the manufacturer of the binder discloses information on the recommended production temperature; this method has been used for modified and special binders.
- Variant 3—the viscosity range of the binder is used as the basis for the independent determination of the production temperature; in this case the viscosity–temperature relationship should be defined to allow determination of the range of temperatures that produces the needed binder viscosity.

An overview of SMA production temperatures according to selected documents is displayed in Table 9.1. A wide range of temperatures is specified in this table. One should remember that each increase of the mixing temperature enhances the risk of binder–mastic draining off the aggregate while also increasing the binder aging. The classic draindown test with Schellenberg's method is carried out at 170°C, producing incomplete information about its behavior at higher temperatures. That is why it is a good idea to conduct another draindown test at a higher temperature that reflects the possible SMA production temperature (see Chapter 8).

9.3.2 Mixing: Basic Information

Mixing the components of a bituminous mixture, proportioned by weight from hot-bins, in a batch asphalt plant takes place in a pugmill. Contemporary batch plants have pugmills of various sizes, usually from 1 ton to 8 tons. Despite the different sizes of pugmills and the resulting output of the plants, mixing time, by and large, remains at the same level for all plants (USACE Handbook, 2000). Determining the suitable amounts of materials to batch given the pugmill's volume is quite a significant step

TABLE 9.1
Recommended Maximum Production Temperatures of an SMA Mixture for Example Binders according to Various Regulations

	Maximum Mixture Temperature in Asphalt Plant (°C)		
Type of Binder	Austria ONORM B 3584 (Table 13)	Germany ZTV Asphalt-StB 07 (Table 5)	EN 13108-5:2006 (Table 16)
Pen grade 50/70	<190	<190	<190
Pen grade 70/100	<180	<180	<180
Pen grade 160/220	—	—	<170
Modified	<190 (PMB 45/80–x) or <200 (PMB 25/55–x)	<180 (PMB 25/55–55)	According to data from PMB producer

Note: PMB = Polymer modified binder.

and is one of the decisive actions undertaken during the plant's calibration. The quantity of material intended for mixing in one cycle may neither be too large nor too small in comparison with the pugmill's volume. In the case of an excessive charge of material, the mixing will be ineffective; the mixture will remain partially unmixed and the stabilizer will not be distributed throughout the mixture. An insufficient amount of material in the pugmill will result in throwing the mixture out of the mixing chamber instead of mixing it properly, an acceleration of the binder-aging process and, again, the potential of destroying the stabilizer.

In a drum-mix plant, components are continually delivered into a constantly mixing drum. Thus the control of the constituents' proportions is exercised through the adjustment of the batching rate. The mixing time depends on the shifting rate of materials inside the drum, which can be affected by a variety of factors such as the length of the drum and the angle of the drum.

9.3.3 SEQUENCE OF MIXING CONSTITUENTS IN A BATCH ASPHALT PLANT

Establishing the sequence of putting materials into the pugmill is chiefly aimed at securing the final homogeneity of a mixture. Mixing constituents consists of the following two stages:

- Dry mixing—this starts the moment the aggregate is deposited into the pugmill, and it ends the moment the binder batching starts.
- Wet mixing—this starts the moment the binder batching begins, and it ends the moment the mixture is discharged into a trolley delivering hot material to a silo (the pugmill's opening).

With regards to the aforementioned stages of mixing, the following universal principles are well-known:

- Dry mixing
  - The dry mixing time should be limited to a minimum since mixing without binder substantially increases the rate of wear on the pugmill and paddles and promotes the breaking of the weaker grains of an aggregate.
  - Increasing the dry mixing time lowers the output of an asphalt plant.
  - An excessive extension of the dry mixing time with a stabilizer in a loose form may cause its destruction (pulverization) or grinding down to a filler shape.
  - The order of aggregate batching to the pugmill and the moment of the filler delivery has a significant influence on the mixture durability (see Section 9.3.3.3.).
- Wet mixing
  - An excessive extension of the wet mixing time causes a higher aging rate of the binder.
  - Despite a proper mixing time, a granulated stabilizer may not be well-dispersed in the mixture; this can be caused by a poor quality stabilizer, so, it is worthwhile to occasionally check its quality.

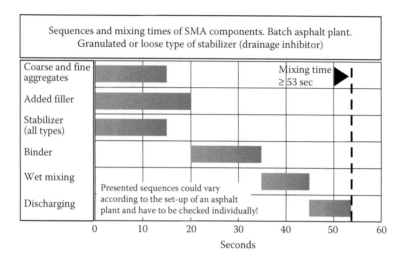

FIGURE 9.2   The batching sequence in a batch-type asphalt plant according to German DAV handbook. (From Drüschner, L. and Schäfer, V., Splittmastixasphalt. DAV Leitfaden. Deutscher Asphaltverband, 2000. With permission.)

According to the German DAV handbook [Drüschner and Schäffer, 2000], the batching sequence does not depend on the kind of stabilizer. There is an assumption in the DAV handbook that the total mixing time of a cycle should be longer than 53 seconds and consist of the actions shown in Figure 9.2. However, depending on the form of stabilizer (loose fibers or granules), various batching patterns have been cited in other publications, including German ones (Graf, 2006; Schünemann, 2007). These cases are discussed in Sections 9.3.3.1 and 9.3.3.2.

Aside from the universal procedure according to the DAV handbook, there are various batching patterns, depending on the form of stabilizer (loose or granulated fibers), established and practiced by many producers of SMA mixtures.

9.3.3.1   Mixing SMA with a Granulated Stabilizer

There is a widespread opinion among a large body of practitioners in the field of SMA production that, after all, the sequence of mixing should vary, depending on the kind of stabilizer. In the case of a granulated stabilizer, this is typically batched with the filler and then mixed with the aggregate without any special extra time for dry mixing (Figure 9.3). The extension of mixing time comes after the binder injection into the pugmill, where an additional 10 seconds of wet mixing time is provided (Graf, 2006). In Figure 9.4, another approach to the mixing sequences of granulated stabilizer is shown.

Although some claim that the use of granulated stabilizers does not involve the extension of the dry mixing time, we might—depending on job site results (i.e., fat spots)—be faced with the necessity of such an action to ensure that the granulate has been fully disintegrated.

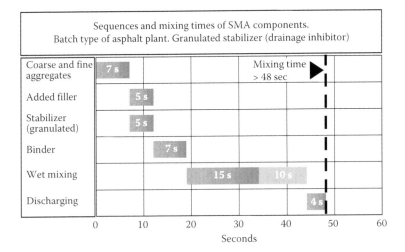

FIGURE 9.3   The batching sequence of SMA mixture constituents into a pugmill with the use of a granulated cellulose fiber stabilizer. Notice that time depends on the type of pugmill. (From Graf, K., Splittmastixasphalt - Anwendung und Bewährung. Rettenmaier Seminar eSeMA'06. Zakopane [Poland], 2006. With permission.)

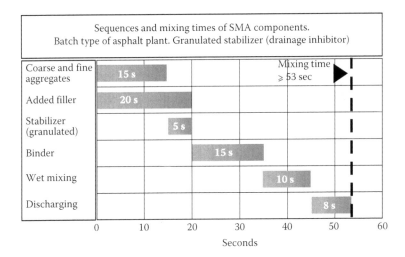

FIGURE 9.4   Batching sequence of SMA mixture constituents into a pugmill with the use of a granulated cellulose fiber stabilizer. Notice that time depends on the type of pugmill. (From Schünemann, M., Faserqualität. Eine wesentliche Voraussetzung zum Herstellen von qualitätsgerechten Asphaltbefestigungen. Rettenmaier Seminar eSeMA'07, Zakopane [Poland], 2007.)

9.3.3.2 Mixing SMA with a Loose Stabilizer

When incorporating a loose stabilizer, which is usually packed in shrink-wrapped bags, extra time is needed for dry mixing the stabilizer with aggregate. A bag of stabilizer is thrown into the pugmill when the filler is being batched, and then an extra period (about 3 seconds) of dry mixing the fibers with the aggregate follows. Because of this, they are released from the bag and evenly distributed in the mixture. The binder is batched onto the dispersed fibers and aggregates, and additional wet stage mixing time follows. Again, one should to remember that too long a dry mixing time could destroy loose fibers.

Figure 9.5 depicts one sequence of the batching of SMA constituents into the pugmill with the use of a loose stabilizer. An example of another batching sequence is shown in Figure 9.6.

The NAPA SMA Guidelines QIS 122 recommend increasing the dry mixing time by 5–15 seconds in comparison with other mixtures without stabilizers. The wet mixing time should be increased by a minimum of 5 seconds when cellulose fibers are used and by not more than 5 seconds when mineral fibers are used.

9.3.3.3 Other Patterns of Batching Constituents in a Batch Plant

The KGO-III (Viman et al., 2004) method has been experimentally applied in Sweden since 2002. It consists of changing the order of batching and mixing the constituents in a batch plant. The suggested KGO-III mixing order is as follows:

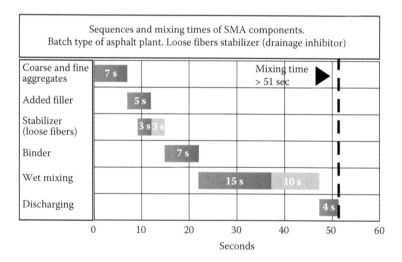

FIGURE 9.5 One sequence of batching SMA constituents into a pugmill using a loose form of stabilizer. Notice that the times depend on the type of pugmill. (From Graf, K., Splittmastixasphalt - Anwendung und Bewährung. Rettenmaier Seminar eSeMA'06. Zakopane [Poland], 2006. With permission.)

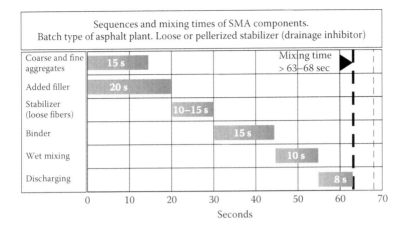

Sequences and mixing times of SMA components.
Batch type of asphalt plant. Loose or pellerized stabilizer (drainage inhibitor)

FIGURE 9.6   Another sequence of batching SMA constituents into a pugmill using a loose form of stabilizer. Notice that the times depend on the type of pugmill. (From Schünemann, M., Faserqualität. Eine wesentliche Voraussetzung zum Herstellen von qualitätsgerechten Asphaltbefestigungen. Rettenmaier Seminar eSeMA'07, Zakopane [Poland], 2007.)

- Stage 1—mixing binder with an aggregate larger than 4 mm
- Stage 2—adding filler only to dissolve it in the mixture
- Stage 3—adding fine fractions (0.063–4 mm)

Changing the order of batching and mixing are aimed at achieving a thicker binder film on the coarse aggregate. As we know, the smallest grains of aggregate are the first to be coated with binder, in a way capturing the binder and interfering with the creation of thick binder films on the coarse aggregate. Therefore the first stage of the KGO-III approach is intended to coat the bigger grains before the smaller ones.

According to Viman et al. (2004), by employing this approach, about 0.5% less binder may be used in comparison with a typical mixture. It has also been shown that the production temperature of the mixture can be reduced by approximately 30°C. Thus not only can the manufacturer benefit from savings in binder and energy costs, but the process is also favorable for the environment due to lower air pollution (e.g., less odors, fumes, and heat). Six different plants and manufacturers have produced in total 500,000 tons of the mixture according to KGO-III since 1998.

9.3.4   PRODUCTION OF SMA IN A DRUM-MIX PLANT

There are two types of drum-mix plants: parallel-flow drum-mix plants and counter-flow drum-mix plants (conventional and double barrel). In drum-mix plants, cool aggregate is delivered from cold-feed bins to a dryer–mixer and then into a silo through a slat conveyor. The mixture gradation control is exercised through establishing suitable proportions of individual aggregate fractions in cold-feed bins and the rate of aggregate supplied by the feeder belt. The conveyor is equipped with a weight and speed-control system that enables control over the coating plant's throughput in tons per hour. The general categories of drum-mix plants depend on the flow direction of the aggregate relative to the hot air movement from the burner. In parallel-flow drum-mix plants, the aggregate and hot air move in the same direction, while in counter-flow drum-mix plants, they move in opposite directions.

In classic parallel-flow drum-mix plants, the asphalt binder is delivered to a dryer–mixer and injected on the aggregate tumbling inside, which poses a risk of direct contact between the binder and exhaust gases from the dryer's burner. That is why various solutions have been adopted that place the binder batching point away from the burner and the aggregate drying zone.

In counter-flow drum-mix plants, the aggregate moves in the direction opposite to the movement of exhaust gases. Many precautionary design measures have been built in and are supposed to provide a significant reduction in emissions, a reduction in the exhaust gas temperature, and protection of the binder against overheating. A lengthened part of the drum where mixing takes place, an extra coater, and an embedded burner are a few examples of these solutions. In counter-flow drum-mix plants of the double-barrel type, the aggregate is dried in an inner drum, then discharged into the surrounding outer drum, where it is mixed with binder while being protected from the burner's high temperature. The final asphalt mix is transported to a storage silo with a conveying device.

The SMA aggregate mix contains more than 70% m/m coarse particles, which means that a substantially higher amount of energy is needed to dry and heat them than when manufacturing asphaltic concrete. Furthermore, it may be necessary to reduce the output of the coating plant and to extend the veil of aggregate flowing through the dryer–mixer per time unit to obtain the proper aggregate temperature.

Adding a loose fiber stabilizer represents a severe handicap from a production control standpoint; the location of batching should be chosen carefully so that the loose stabilizer cannot be captured by hot exhaust gases. However, the application of granulated fibers does not present major problems, provided that the point of their addition into the mix is properly determined. Adding granulate to the drum behind the burner (parallel-flow or conventional counter-flow) or to the outer drum before adding binder (counter-flow double-barrel system) is the rule.

9.3.5 Fluctuations in Batching of Components

SMA mixtures are very susceptible to overbatching of the binder quantity during production. When this occurs, an extra amount of binder appearing in a mixture is followed by the rapid decrease of mastic consistency and a change in the volume ratios of the entire SMA. Moreover the risk of binder or mastic draindown becomes much higher.

The strength of SMA is founded on its mineral skeleton. So the properties of SMA are dependent on the gradation and, consequently, are extremely susceptible to the shifting of the gradation curve. As discussed in Chapter 6, we should consider the impact of the change of coarse aggregate content on the breakpoint (BP) sieve. Another way to express that concept is that a substantial change in the content of the largest coarse aggregates causes such significant changes to the volume relations of a mixture that they can greatly diminish an SMA's strengths.

9.3.6 SYSTEMS OF BATCHING STABILIZERS

Modern asphalt plants are equipped with integrated silos—delivering and weighing systems mostly designed for granulated or pelletized stabilizers. These stabilizers are stored in silos and then transferred to the scales, proportioned by weight or by volume and blown by compressed air into the pugmill (Figure 9.7). In older plants,

FIGURE 9.7 A system for batching granulated fibers into mixing plants. The photograph shows an open silo for granulates ready for loading. (Photo courtesy of Bartosz Wojczakowski.)

more or less sophisticated batching methods (i.e., proportioning by weight or by volume) are applied to weigh out or measure out a batch of granulated or loose fibers directly into the pugmill (Figure 9.8).

Automatic batching devices for nongranulated (loose fibers) are rare. Bags of loose fibers are emptied through a special charging box into the pugmill (Figure 9.9) or directly through an opening in the upper part of the pugmill's cover.

FIGURE 9.8 Manual proportioning control of a granulated stabilizer directly into the pugmill. (Photo courtesy of Bartosz Wojczakowski.)

FIGURE 9.9 Manual charging box with a motor-driven chute. (Photo courtesy of Bartosz Wojczakowski.)

When there are no bins for stabilizer, an effective solution may be the direct delivery of granulated cellulose fibers from a tanker into the pugmill.

9.4   STORING THE READY-MADE SMA MIXTURE IN A SILO

Most guidelines for SMA do not recommend producing large quantities of a mixture and holding it in reserve for too long for future use. SMA stockpiling is an uncertain business due to the risk of binder draindown when storing hot mixture. Such limitations cause problems when the asphalt plant has a low output and large amounts of mixture are needed to be supplied to a laydown site. It can be assumed that, depending on the temperature, mineral composition, and binder content, the mixture may be stored for up to 2 or 3 hours at a high temperature. The risk of draindown is magnified by the SMA's extended storage time in a silo, which is why it should be sent to the work site within a reasonable amount of time after production.

It is important to monitor the decreasing temperature of a mixture in a silo. Not every silo is equipped with heated walls or even a heated chute. Allowing the mix to cool substantially in the silo may cause many problems, especially with a polymer-modified binder. It has been stated in the U.S. Department of Air Force guidelines (ETL 04-8) that a ready-made SMA mixture can be stored for no longer than 1 hour in uninsulated silos or 4 hours in thermally insulated silos. If, for any reason, the hot SMA has to be stored longer in a silo, then it is worthwhile to incorporate a higher amount of stabilizer.

9.5   SMA MIXTURE PRODUCTION CONTROL

The SMA mixture production process, like that of other asphalt mixtures, is subject to procedures to control its mixed components and other selected properties. These procedures differ among various countries, but in almost all cases, the control of gradation and binder content form their common root.

The control of the production process mostly consists of periodically checking the components of the produced mixture in relation to the approved laboratory recipe. Generally, two methods applied world wide may be distinguished as follows:

• The control of the aggregate gradation on selected sieves and the content of soluble binder within given tolerances of production accuracy
• The control of SMA volume properties (e.g., the content of voids in a mineral aggregate (VMA) and the content of voids in a compacted asphalt mixture)

It is good to emphasize the control of the volumetric properties during the production process; it is even more logical to verify the volume ratios determined at the design stage during production. Furthermore, control of the aggregate gradation alone, within allowable tolerances, does not guarantee the correct volume ratio of coarse aggregates, mastic, and air voids.

9.5.1 Control according to the German Document ZTV Asphalt-StB 07

The quality of the produced mixture may be determined by ZTV Asphalt-StB 07. The following tests are included in the scope of methodical control activities (Table 26 ZTV Asphalt-StB 07):

- Gradation on selected sieves
- Soluble binder content
- R&B softening point of the recovered binder
- Bulk density and air-void content in compacted Marshall samples

The factory production control (FPC) organization and the whole verification of a conformity system* 2 + is in accordance with requirements contained within the European Standard EN 13108-21.

9.5.2 Control according to U.S. Documents

The distinctive feature of the contemporary U.S. approach to production control is the application of statistical methods and fines as punishment for exceeding admissible limits or deviations from the mix design. (Sometimes bonuses, or incentives, are also used to reward exceptional production consistency, but penalties are more common.)

The quality control methods widely employed in the United States are called quality control/quality assurance (QC/QA). These systems typically involve testing recently produced bituminous mixtures for the following (USACE Handbook, 2000):

- Mix components—binder content and gradation of the mineral aggregate
- Physical properties of the mixture measured on compacted Marshall or gyratory samples
  - Air-void content
  - VMA
  - Voids filled with binder (VFB)
  - Density
  - Stability and flow, if required

Under QC/QA specifications, the mixture manufacturer is charged with the responsibility for quality production control (QC). Its laboratory's duty is to undertake control activities. Verifying the operation of QC, or QA, is the investor's (or owner's)

---

* The number 2+ is a designation of the level of attestation of conformity (AoC) system according to Annex III of the Construction Products Directive (Council Directive 89/106/EEC). The system describe the tasks of the construction product manufacturer (here, the asphalt mix producer) and the tasks for the notified body (control organization). In system 2+, which is the most often used for construction products, the manufacturer should carry out the initial type-testing of the product (e.g., test for conformity of the mixture with the specification), establish FPC system and, possibly, testing samples taken at the asphalt plant in accordance with a prescribed test plan.

responsibility. It is necessary to stress again that in the United States sliding scales of fines are imposed, depending on the statistical results of production control.

Every producer of a mixture must formulate and submit for approval a quality control plan (QC plan). Broadly speaking, this type of plan* comprises such details as the determined features under control and the frequency of their testing, the methods for recording results, and the corrective actions taken in case of excessive deviations from a formula. Additionally, the QC plan outlines the number and the frequency of equipment inspections, the calibration of instruments, and document management. More information on control using QC/QA specifications may be found in various U.S. publications (*The Asphalt Handbook*, 1989; USACE Handbook, 2000).

9.5.3   FACTORY PRODUCTION CONTROL AFTER EN 13108-21

Issues concerning procedures for production of bituminous mixtures under EN 13108-21 and specified tolerances are discussed in Chapter 14.

9.5.4   OTHER EXAMPLES OF PRODUCTION CONTROL

Some brief examples from around the world include

- In some countries, the accuracy of stabilizer dosage is ±10% of its mass as determined in a laboratory formula.
- If binder viscosity is the basis for establishing the mixing temperature of a binder and aggregate, some specifications may quote a permissible deviation from that temperature (e.g., ±10°C).
- In most countries, the permissible deviation of a mixture gradation and the content of binder depends on the number of tested samples.
- Some regulations provide clauses permitting a situation in which the gradation curve of a produced mix, considering given deviations, may deviate outside the area of the gradation envelope (between the so-called upper and lower limit curves).

9.5.5   PROBLEMS AND TIPS

Today almost all asphalt plants are computer controlled. It is worthwhile mentioning the necessity for caution during SMA production. This also applies to these modern asphalt plants. Usually such machines give personnel a feeling of confidence and a sense of complete control over the production process. However, many instances have proved that an excess of trust leads to problems (some of which are described in Chapter 11).

Particular attention should be paid to checking weight batchers, thermometers, and proportioning meters. All data produced by the computer system should be periodically checked. A good example of the type of problems that may occur is the clogging of batching devices for granulated stabilizers between the balance and the chute

---

* These terms and range of activities have also been described in the European standard EN 13108-21.

FIGURE 9.10 Performance inspection of the granulated stabilizer batching into a plant's pugmill. (Photo courtesy of Krzysztof Błażejowski.)

FIGURE 9.11 The SMA asphalt mixture with the stabilizer after extraction and before aggregate screening. (Photo courtesy of Krzysztof Błażejowski.)

gate. The actual amount of stabilizer weighed out by the machine can be checked by disconnecting the chute pipe and measuring the amount of material batched during a production cycle (Figure 9.10). Another typical method is performing an extraction and checking for the presence of the stabilizer in the mixture (Figure 9.11). Unfortunately this method does not allow for the determination of the exact amounts of fibers added to each batch.

Finally, a brief note about extraction. While performing an extraction of an SMA mixture, the rules for the correct preparation of the sample should be observed (having the stabilizer in mind as the mixture's constituent—see Figure 9.11). The stabilizer fibers should be removed prior to starting the aggregate screening.

10 Transport and Laydown of the SMA Mixture

Once the SMA has been produced at a hot mix plant, it must be transported to the job site where it will be placed and compacted. This chapter describes issues that need to be considered during this stage of the construction process.

10.1 TRANSPORT OF THE MIXTURE TO A LAYDOWN SITE

While it may seem that hauling SMA to the job site is straightforward, there are things that can go wrong that can affect the rest of the construction process and ultimately the performance of the pavement. This section will describe potential problems and how to prevent them.

10.1.1 LOADING A TRUCK

The hot SMA mixture, at first stored in a silo, must be transported to the work site. The way the mixture is loaded into the bed of the truck affects its subsequent quality. The following rules should be observed to prevent segregation of the mix:

- The discharge into the bed of the truck should be performed in defined drops; generally the first drop should be placed in the front part of the load bed, the second one toward the rear, and the rest evenly distributed in the middle (Hensley, undated; Roberts et al., 1996).
- The addition of small amounts of mixture to reach the full weight limit of a loaded truck may bring about the separation of larger particles of the aggregate from the mixture (segregation) and should be avoided.
- When using trucks with large beds, the truck should be loaded in four or more separate drops, overlapping the individual dropped piles to help remix the material.

Detailed instructions for loading mixtures out of silos may be found in the USACE Handbook (2000).

The draindown effect (i.e., binder or mastic draindown followed by its collection on the bottom) may happen under the following circumstances (Ulmgren, 2000):

- When there is excessive binder and mastic
- When the temperature of the mixture is too high
- When there is insufficient stabilizer or it is of poor quality

Also, remember to suitably prepare the truck bed prior to loading the mixture. The bed should be clean and free from dirt (see requirements according to EN 13108-21). Sideboards and bottoms of the beds of the delivery trucks should be covered with a special release agent (antiadhesive fluid) to avoid adhesion; diesel oil and other fluxing agents that may degrade the bituminous binder are not permitted to be used in this capacity. The truck bed should be free of major dents in which substantial amounts of release agents could gather. The boards should be evenly coated, and any excess release agent that may collect in deformations of the bed should be removed prior to loading the truck.

10.1.2 TRANSPORT

The main objective of transporting a mixture from the asphalt plant to the laydown site is to deliver the mix without changing its properties and in a state that allows for appropriate placement and compaction. Minimizing heat loss and preventing segregation of the mixture constituents are two of the most important issues regarding transportation. Requirements for transporting the mixture to the work site are often defined in technical specifications. Typical requirements are shown, as examples, in the following:

- Specified maximum distance between the asphalt plant and laydown site (e.g., 40 km)
- Maximum travel time from the asphalt plant to the work site (e.g., 2 hours)

Specifying the distance between the asphalt plant and the construction site can be quite a misleading operation since traveling a distance of 40 km in a rural area does not take the same amount of time as driving the same distance in a large city. Driving time is the more accurate requirement; however, some differences may be noted even in that. Mix will not cool as much after 2 hours during the summer as it will after the same amount of time in the late autumn. Probably the best way of setting the requirements for transportation is to establish temperature conditions for the mixture. This method of specifying requirements may be found in the European standards concerning bituminous mixtures (e.g., in EN 13108-5 [the European standard on SMA]). There are maximum production temperatures specified for various types of bituminous binders and a minimum supply temperature for the delivered mixture. One should meet that temperature range when planning the production and transport of a mixture to a work site, taking into account such factors as the prevailing weather conditions, and the location of the asphalt plant. This is definitely the more common solution.

When transporting hot SMA, a crust of cooler mixture is formed on its surface. The higher the degree of cooling of the mixture, the thicker the crust, and the more problems it causes. As long as the mixture is well-protected thermally (by a tarp or insulation), the layer of cooler mixture will be thin and the chance to intermix it with the rest of the material during laydown will remain high. Significant cooling of the mixture during transportation or while waiting for discharge leads to the build-up of a pretty thick crust. Large pieces of cool, unmixed material cause a nonuniform texture in a laid-down course (for details see Chapter 11). In that case, good mixing may be provided by a buffer like a material transfer vehicle (MTV) or Shuttle Buggy (Brown, 2002) (see Section 10.1.4).

There is an analysis concerning the heat losses of an asphalt mix during site transportation in the paper by Spuziak (2002). The analysis assumes that heat losses during mixture transportation may range from 5°C/hr (with good insulation) up to 48°C/hr (in exposed locations). The drop in the mixture temperature depends on the following:

- Weather conditions during transportation, including the air temperature, humidity, and wind velocity
- Air streamline speed (the speed of the truck loaded with mixture)
- Time of the mixture haulage
- Mass of the load
- Shape of the truck bed and its insulation properties

Concerning weather conditions, wind velocity is a decisive factor in the rate of cooling of the mixture during transportation. The mass, volume, and shape of the hauled material is also of great significance. Any reduction in the thickness of the transported layer of mixture induces quicker cooling, hence the shape of the truck bed is also of considerable importance. Taking this into account, beds with rounded or wedged edges are ideal (Figure 10.1) (Ulmgren, 2000, Spuziak, 2002).

In cool weather, the best way to retain the mixture temperature is to use an appropriate means of transport. Trucks with insulated beds are the best solution, then the effect of the reduction of heat waste is perfectly obvious (Figures 10.2 and 10.3). In the case of such a truck being unavailable, tight and well-fitting tarps on trucks should be adopted as a minimum. Minimizing drops in the temperatures of SMA mixtures containing polymer modified binder (PMB) is particularly important.

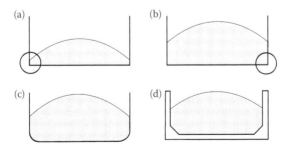

FIGURE 10.1 Location of cooled spots of a mixture on a cross section of a truck bed (a) and (b). Truck bed shape prevents cold corners (c) and (d). (From Spuziak, W., *Proceedings of the 6th International Conference Durable and Safe Road Pavements*, Kielce, Poland, 2000. With permission.)

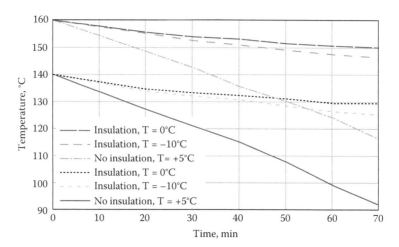

FIGURE 10.2 The drop in temperature of mixtures with and without insulation over given periods. In these scenarios, the ambient temperature is −10°C, 0°C, or + 5°C; the bulk mixture is hauled in the truck bed or in a special insulated container; and the mixture temperature at the plant is either 140°C or 160°C. (From Spuziak, W., *Proceedings of the 6th International Conference Durable and Safe Road Pavements*, Kielce, Poland, 2000. With permission.)

10.1.3 Discharging

The suitable discharge of a mixture from a truck to the paver hopper helps to avoid some problems. The dump truck with the mixture should pull in directly in front of the paver to allow for the following:

- The truck to be positioned exactly on its motion axis, enabling the push rollers on the paver to contact both rear wheels of the truck.
- The truck can avoid bumping into the paver; bumping the paver can cause a bump in the pavement.

As a general rule, the paver should approach the truck and not the other way round (then the paver can push the dump truck without bumping).

10.1.4 Other Techniques for Discharging Mixture

Special self-propelled hoppers (e.g., MTVs) have been used to avoid segregation during the discharge of a mixture from the dump truck to the paver hopper and to minimize the risk of placing cool mixture (by remixing). These mobile machines operate between the dump truck and the paver to eliminate any contact between the two. MTVs are equipped with conveyors to transport the mixture to the paver hopper. These devices eliminate jarring of the paver caused by contact with the delivery trucks or by the mass of mix dropping into the hopper. They hold a larger mass of material than a paver hopper, which helps to keep the temperature high, and allows the paver to keep moving forward. Another advantage of MTVs is the possibility of remixing the mixture constituents. This eliminates the risk of segregation and

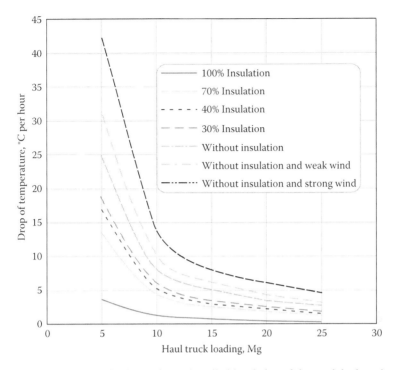

FIGURE 10.3 Impact of load capacity and applied insulation of the truck bed on the temperature drop of mixture in various weather conditions. (From Spuziak, W., *Proceedings of the 6th International Conference Durable and Safe Road Pavements*, Kielce, Poland, 2000. With permission.)

enables the breaking of large lumps of cool mixture that may occur during transportation. Then the mixture is transferred to the paver hopper using a conveyor belt. MTVs are especially useful when laying mixtures susceptible to segregation and paving during cooler periods of the year (see Chapter 11, Section 11.8).

10.2 LAYDOWN CONDITIONS

Due to the resistance to compaction provided by the coarse aggregate skeleton, SMA mixtures can cause problems during laydown. Thus in almost all countries that have adopted SMA, some additional requirements on conditions at the work site must be observed, including the appropriate temperature, wind velocity, and the absence of rain.

10.2.1 Minimum Temperature of Air and Underlying Layer

To ensure that weather conditions are acceptable during laydown, the air temperature may be checked often. In many countries, however, air temperature is not as important as the temperature of the course on which the hot mixture is going to be placed. Required laydown conditions frequently hinge on the thickness of the constructed layer, because thick layers cool more slowly, and on the type of binder used, whether modified or unmodified.

Two types of weather conditions may be distinguished in specifications: conditions at the time of laying and conditions within 24 hours prior to starting work. Average requirements for the minimum air temperature at the time of placing an SMA mixture range from 5°C to 10°C as exemplified by the following:

- German Dav documents stipulate an air temperature of not less than +5°C (Milster et al., 2004)
- The Czech standard CSN 736121 requires +5°C.
- The Polish guidelines WT2 2008 specify +10°C for courses less than 3 cm thick and +5°C for courses greater than or equal to 3 cm thick.
- The Australian NAS AAPA 2004 calls for +5°C, with the limitation that laying courses thinner than 4 cm or containing a modified binder should be avoided at such low temperatures.

In general, the temperature requirements are becoming stricter (increased) in cases where the thickness of a course is less than 40 mm; for that thickness, a minimum requirement of +10°C is commonly specified. In Germany's ZTV Asphalt-StB 07, the weather limitations for SMA laying are as follows:

- When the course thickness is less than 3 cm, the minimum air temperature is +10°C, and the minimum sublayer temperature is +5°C.
- When the course thickness is more than 3 cm, the minimum air temperature is 5°C.

The average requirements for the specified underlying layer's temperatures are similar and are also within a range of 5–10°C. A higher value is often used for mixtures with modified binder. Another document, the U.S. Federal Aviation Administration Advisory Circular 150/5370-10B on civil airfields, has stipulated the following minimum temperatures of the underlying course during laydown, which depend on the thickness of the asphalt course being placed:

- For a course less than 2.5 cm thick, the underlying layer temperature must not be lower than 10°C.
- For a course 2.5–7.5 cm thick, the underlying layer temperature must not be lower than 7°C.
- For a course greater than 7.5 cm thick, the underlying layer temperature must not be lower than 4°C

Slightly lower values than those just cited have been adopted when specifying temperatures during the day directly preceding the start of laying down the hot mixture. For example, according to the Czech standard CSN 73 6121, the air temperature must be no less than 3°C (5°C for a thin course), and according to the Polish guidelines WT2 2008, it must be at least 0°C (5°C for a thin course).

Another, indirect method of specifying temperature requirements is by limiting the permissible calendar period of SMA laydown—for example, by imposing a ban on works over a specified period of the year (e.g., between October 15 and March 15).

Finally, it is worth noting that a bituminous mixture with an unmodified binder can be laid down at a temperature slightly lower than a bituminous mixture with a modified binder.

10.2.2 UNDERLYING LAYER'S MAXIMUM TEMPERATURE

The limitation of minimum temperatures of SMA laydown has been commonly understood and adopted; however, it is quite a different matter when laying down the SMA mixture on an underlying layer that has an excessive temperature, as in the following examples:

- On a recently placed intermediate course that has not yet cooled
- While *hot recycling in situ* (or *in place*) with the simultaneous placement of a new wearing course, so-called *hot remix plus*
- During the execution of the so-called *Kompaktasphalt*—the laydown and compaction of two courses (intermediate and wearing) at the same time with one passage of a paver (see Section 10.5).

As a general rule (see Section 10.4), rollers compacting an SMA layer operate directly behind the paver. This practice is correct in typical circumstances since the mixture cools quickly. But when the underlying layer is warm, such a method may create some problems, with the most frequent one resulting in the binder bleeding under the rollers and creating fat spots.

The following comments and observations deserve mention:

- The rolling operation cannot be started until the spread mixture has reached the optimum start temperature related to the viscosity of the applied binder; therefore placing the mixture on an underlying course that is still warm—with temperature exceeding 50–70°C (e.g., on a course made several hours earlier)—should be accompanied by ongoing temperature control.
- Because of the warm underlying course, the SMA does not cool as fast, consequently giving more time for compacting and enabling a greater distance between the rollers and the paver.
- In cases of fat spots being squeezed out by the rollers, rolling should be stopped until the mixture temperature in the course drops to a point where rolling does not overcompact the mat; the time (or temperature) at which rolling can resume should be determined experimentally for specific binders.

Unfavorable air temperatures (too low) and an excessively warm underlying course cause various problems. Having an excessively warm underlying layer creates more favorable laydown conditions because this allows for more time to work with the mix.

10.2.3 WIND AND RAIN

There are additional requirements regarding issues such as wind speed at the laydown site in many national specifications. Let us take the Polish guidelines WT2 2008, for instance, in which the maximum allowable wind speed has been set at 16 m/sec (58 km/hr). The values of wind speed at the construction site are undoubtedly a factor of tremendous significance, especially when placing thin courses of SMA.

Frequent rainfall and the necessity of placing an SMA on a wet underlying layer are problems that often beset road builders. Most regulations do not allow SMA laydown on a wet surface or during rainfall (Figure 10.4). The layer of water covering the surface of the underlying layer causes a rapid drop in the SMA mixture temperature and also counteracts the formation of durable interlayer bonding. It has been estimated that a layer of

FIGURE 10.4   Laying down SMA during rainfall is not a desirable (and is mostly a forbidden) practice. (Photo courtesy of Krzysztof Błażejowski.)

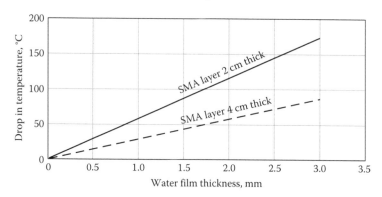

FIGURE 10.5   Influence of water film on the drop in SMA mixture temperature while compacting on a wet surface. (From Drüschner, L., *Asphalt Review*, 25, 1, 2006. With permission; Schellenberger, W., Einbau von Asphaltbeton und der Einfluss der Witterungsbedingungen. Asphalt 7–8, 1997. With permission.)

water about 1 mm thick causes a drop in the temperature of a 4-cm thick SMA course of about 30°C (Drüschner, 2006). All in all, the thinner the SMA course being placed on a wet surface, the faster the mixture gets cold. Figure 10.5 shows the relationship between the amount of water on a surface and the drop in the mixture temperature.

Obviously a mixture should not be laid on a frozen surface or on one covered with snow or ice. But keep in mind that, when justified, the supervising engineer may approve carrying out works in unfavorable conditions, provided that all requirements with reference to the achieved properties of a course are satisfied. Typical technical regulations have stipulated that the laydown of an asphalt mixture in adverse conditions might be permitted as long as the following are true:

- All final requirements for the course are fulfilled.
- Good bonding of structural and wearing courses is secured.
- Good performance of construction joints is guaranteed.

10.2.4 SURFACE PREPARATION

The rule for SMA surface preparation is the same that as for other binder courses. The surface of the underlying course should be even, clean, and free from dirt that could disturb the appropriate bonding of the laid-down mixture with the layer underneath. Any soiling of the surface (Figure 10.6) should be removed from the pavement. Oil and fuel stains should be cleaned with an absorbent material, and their residue should be washed off. The remains of an old mixture atop an underlying layer should be removed because they may cause partial unevenness of the spread SMA mixture (see Figure 11.33). When placing an SMA course on an old pavement, one should see that patches made of mastic asphalt have been removed and a new patch of asphalt concrete has been made. Leaving the mastic asphalt in place runs the risk of bleeding the binder from the patch onto the surface of the SMA course. Potholes and cracks should also be filled and sealed prior to placing the next course.

FIGURE 10.6 Soiling of an intermediate course just before laying an SMA mixture. (Photo courtesy of Krzysztof Błażejowski.)

10.2.4.1 Bonding Layer (Tack Coat)

Attention should be paid to the proper laydown of the tack coat beneath the SMA course. When the surface is uneven, with pockets collecting the binder of the tack coat, its excess should be completely removed, otherwise it is likely that fat spots may appear on the surface of the SMA course. The tack coat underneath the SMA is necessary, but the amount and type of required binder (bituminous emulsion or cutback binder) should be suitably estimated, depending on the state and type of surface. After all, we do not need a tack coat that is too thick or too soft because it can become a failure (sliding) plane under the SMA. A special binder (e.g., a special bituminous emulsion) should be used for interlayer bonding. Except for special cases, typical binder emulsions made of soft binder (e.g., 160/220 [Pen@25]) are not recommended. An average amount of binder (residual) for tack coating under an SMA varies within the range of 0.1–0.3 kg/m². For example, in Germany according to ZTV Asphalt-StB 07, the recommended quantities of a polymer modified emulsion C60BP1-S* (former TL PmOB Art C U 60 K) for interlayer bonding for roads with heavy or very heavy traffic depend on the type of layer beneath the SMA as follows:

- Newly constructed asphalt intermediate layer: 150–250 g/m²
- Asphalt intermediate layer with cold milling surface: 250–350 g/m²
- Very porous asphalt intermediate layer or ravelled surface: 250–350 g/m²

Note: the above amounts of bituminous emulsion are NOT residual binder (from the emulsion).

Additionally there is a requirement for the strength of the connection between layers; the splitting force has to be at least 15 kN. A lack of interlayer bonding adversely affects the pavement's service life; it can also cause some negative effects in the form of slippage under a roller (see Figure 11.32).

10.2.5 PREPARING PAVEMENT FACILITIES

The existing pavement facilities should be prepared prior to placing the mixture of a wearing course. This preparation should result in the following:

- Appropriate, tight, and durable bonding between asphalt course and facilities (e.g., with the use of suitable polymer modified asphalt tapes [Figure 10.7])
- Protection against damage or infiltration of excess asphalt mixture (e.g., around a drain or manhole)

10.2.6 PREPARING EDGES OF THE COURSES

The edges of the course should also be prepared. The selected method should ensure durable joints at the transverse and longitudinal edges. Most previous experiences have proved the poor effectiveness of traditional methods like applying hot binder,

---

* Polymer modified emulsion, 60% of binder, rapid type. C60BP1-S is an emulsion designation according to the rules of EN 13808.

(a) (b)

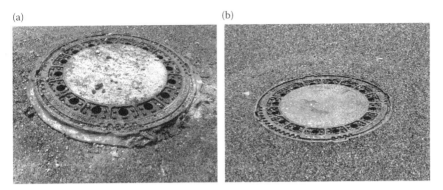

FIGURE 10.7 (a) A drain with PMB tape put on its edges; (b) the same drain after the placement of SMA mixture around it. (Photo courtesy of Krzysztof Błażejowski.)

(a) (b)

FIGURE 10.8 PMB tape for sealing pavement joints (a) when bonding with a layer edge and (b) after application. (Photos courtesy of Śląskie Kruszywa Naturalne sp. z o.o., Poland.)

covering with emulsion, and other similar techniques. As in other cases, PMB tapes (Figure 10.8) put on manually or special PMB compounds spread mechanically (Figure 10.9) have been recommended.

10.3 PLACEMENT OF A MIXTURE

Corrections to mixture problems occurring during delivery of a mixture may be possible up until the moment of placing the SMA. The moment the SMA layer appears behind the screed plate of a paver, the chances of improving the quality diminish to a minimum. After that time only compaction is possible; errors made during the design and manufacturing stages can no longer be corrected.

The vital elements of spreading an SMA layer include the following:

- Appropriate selection of the mixture gradation relative to the layer thickness
- Efficient operation of the paver (when spreading mechanically)
- Suitable manual spreading in places inaccessible to the paver
- Proper use of rollers

(a)   (b)

FIGURE 10.9 PMB compound for sealing pavement joints (a) while laying down with a special device and (b) after application. (Photos courtesy of Śląskie Kruszywa Naturalne sp. z o.o., Poland.)

10.3.1 Layer Thickness

An SMA layer thickness should not be less than three times the maximum aggregate size in the mixture, and in principle, not greater than four times (higher ratios allow better compactability). An appropriately selected layer thickness with regard to gradation enables suitable compaction of the layer (reaching the expected compaction factor). For details on selecting SMA gradation, please refer to Chapter 6.

The layer thickness exerts a considerable influence on the speed of cooling of the layer, eventually involving temperature problems of various types (see Chapter 11). To put it briefly, the thinner the SMA layer, the more difficult the compaction. In addition, a thin layer cools off fast, magnifying compaction problems.

10.3.2 Mixture Temperature during Compaction

Almost all publications on SMA underline the necessity of carefully observing the temperature of the mixture during placement and rolling. The expected range of mixture temperature is determined in different ways; it chiefly depends on the kind of binder, but such factors as the layer thickness and weather conditions are important, too. However, the most important factor is the temperature of the mixture delivered to the construction site and the temperature at the end of effective compaction, below which further rolling becomes ineffective and even harmful.

Minimum temperatures for mixture supplied to a work site according to the European standard EN 13108-5 (which applies only to selected [unmodified] bitumens after EN 12591) are as follows:

- 160°C for paving grade bitumen 40/60
- 150°C for paving grade bitumen 50/70
- 140°C for paving grade bitumen 70/100

In the German DAV SMA handbook (Drüschner and Schäfer, 2000), the aforementioned temperatures are presented in a more general way; the suggested temperature of an SMA mixture in a paver hopper should not be lower than 150°C. The same rule is presented by Bellin (1997).

Different temperatures at the end of the compaction time have been assumed in various publications, from 80–100°C for ordinary binder, and from 120–138°C for modified ones. The U.S. NAPA SMA Guidelines QIS 122 stipulates no rolling when the temperature of a layer drops below 116°C. A temperature of about 100°C has been stated in German documents as the point at which to stop rolling. The minimum temperature at the end of the compacting time may be roughly calculated by adding 50°C to the Ring and Ball (R&B) softening point of the binder used in the mix (Daines, 1985; Read and Whiteoak, 2003).

Other relevant points include the following:

- Problems related to a mix temperature that is either too low or too high are elaborated on in Chapter 11. Additional comments may be found in Section 10.4.2.5, which deals with rolling time.
- Optimum compaction temperatures are related to the viscosity of the added binder. That implies the significance of not only the lower temperature limit of rolling but also the initial temperature of rolling (already described while discussing SMA laydown on a hot underlying layer). A mixture that is too hot also causes problems at placement.
- Remember that spreading mixtures with substantial temperature differences (e.g., from a truck with a hot mixture alternated with another truck with a cool mixture) cause changes in the resistance offered by such mixtures at spreading.
- Appearing here and there in a layer being placed, pieces of a cool mixture may cause the development of an increased content of air voids and hence decrease the pavement's lifespan (Pierce et al., 2002).

All these remarks concerning temperatures do not apply to cases that involve the use of special additives for lowering mixture temperatures that create the so-called warm mixes.

10.3.3 Mechanical Spreading

Nowadays, mechanical placement is the only reliable method of executing an SMA layer. Requirements for pavers can very rarely be found in the specifications.

Basically, selection, setting, and operation of a paver are the responsibilities of the paving contractor. Most road-engineering companies, after gaining experience with SMA pavements, work out their own procedures for spreading, compacting, and achieving the required parameters. The following information might help to establish or improve such procedures.

10.3.3.1 Paver

The appropriate passage of material through the paver plays a key role in the proper spreading of a mixture. After starting in the hopper, the mixture is moved by slat conveyors (with flow gates in older equipment) to augers and then under a screed. During each of these stages the following significant parameters affect the final result:

- The hopper should be fitted with independently lifting wings, and its shape should eliminate places from which the mixture does not slide to the slat conveyor. Such "dead areas" or "cool corners" create accumulations of cool mixture and cause other problems (see Chapter 11). For the same reasons, the insulation of wings is desirable.
- Care should be taken so that the mixture does not adhere to the walls of the hopper where it cools off fairly quickly. These effects can be seen in various forms of segregation (see Chapter 11). It is perhaps worth dedicating one of the paving crew to systematically throw the cooled remains of mixture down in to the middle of the hopper, particularly when work is done on cool and windy days.
- Completely emptying the hopper of mixture should not be permitted. Newly delivered material should be added to the hopper when it is still filled to about 20% of capacity.
- Augers are intended to divide the mixture across the width of the screed plate; the quality of the layer's surface depends, among other things, on appropriate adjustments to the augers. The amount of mixture supplied to the augers should be constant; it can be controlled by setting the slat conveyors' speed and allowing an adequate opening of the flow gates (if applicable).
- The distribution of mixture at the middle of the paver screed plate has a significant influence on the segregation of an SMA mixture. In some machines, there is a feeding screw without a large chain transmission on the axis of the paver; rather the feeding screw is propelled from outside by hydraulic engines and intersecting axis gears.
- The screed plate must be fitted with a heating system. A number of solutions are available and include power supplies, heating fuels, and gas burners. A properly heated screed enables the appropriate travel of the mixture without dragging and pulling out particles.
- Screeds are fitted with vibration systems and rammers. The frequency and amplitude of vibrations and rammers' strokes should be compatibly matched.
- To ensure quality of the placed layer, a constant paver speed should be maintained. Stoppages should be avoided.

10.3.3.2   Other Remarks

It is worth remembering the following when mechanically spreading a layer of SMA mixture:

- Manually scattering mixture over the mechanically placed mixture is not permitted.
- Allowing any vehicles other than rollers on the hot mixture, before it is finally compacted and its temperature drops below the expected level, is not permitted.
- The end of a working lot (the transverse joint) should be finished by cutting; suitable joint bonding ought to be secured prior to laying the next lot—for example, by applying a PMB tape or a special compound (Figures 10.8 and 10.9).
- When executing the SMA layer in separate lanes (i.e., each traffic lane separately), the edge of the first layer should not be cut vertically but with a falling gradient, e.g., of 3:1 (height: width).
- Longitudinal edges should be cut while the mixture is hot. A special cutting wheel installed on a roller is user friendly; it is even more convenient with a metal pusher for the cut mixture located next to the cutting wheel, which eliminates manual removal operations.
- The utmost attention should be paid to the execution of an appropriate and tight longitudinal joint between two SMA layers; it is difficult due to the high content of chippings in mixtures; experiences have proved that the PMB compounds, laid down mechanically, are very effective.

10.3.4   MANUAL PLACEMENT

In general, most regulations and guidelines do not permit manually placing SMA mixtures; however, there are places where there is no other option but to place the mixture manually (e.g., small and irregular pieces of a roadway) (Figure 10.10). In such cases one should remember not to scatter the mixture with shovels but to carry and lay it down. Compaction should be carried out immediately after laying and alignment. Unfortunately such surfaces will differ in structure from adjacent areas spread mechanically, and usually they will be more porous and permeable.

10.4   COMPACTION

Bearing in mind the unique philosophy of designing an SMA aggregate skeleton (see Chapters 2 and 6), SMA compaction may be regarded as the process of forming a structure made up of appropriately arranged and interlocked grains filled with mastic. While rolling, the mixture volume is reduced by the tighter arrangement of particles and a decrease in the content of air voids. The binder makes compacting easier. It plays the role of a lubricant, easing the movement of aggregate at that stage of an SMA's performance.

(a)   (b)

FIGURE 10.10   Spreading the SMA mixture on an exit ramp off a trunk road: (a) manually placing the mixture and (b) compacting edges of a working lot with the use of a vibrating plate. (Photos courtesy of Krzysztof Błażejowski.)

10.4.1   Types and Number of Rollers

The fact remains that SMA is a difficult mixture to compact, therefore the appropriate selection of both the number and type of rollers is of real significance.

10.4.1.1   Types of Rollers

The following types of rollers can be used for compacting SMA mixtures:

- Static—used as basic equipment for SMA compaction. The heavy (finisher) and medium ones operate chiefly in the set. When compacting is executed on thin layers or layers on a stiff underlying, heavy rollers are excluded.
- Vibratory— used for compacting SMA, but only according to some rules mentioned later.

Pneumatic rollers are not typically used due to the risks of mastic sticking to the tires and dragging particles out of the rolled layer and of squeezing mastic out on the surface.

The following rules should be observed when planning for a combination of rollers for a work site:

- The type of rollers should be compatible with the thickness of layers and the ambient conditions.
- The number of rollers should be compatible with the expected area to be paved and the compacting ability of the rollers. At least two or three rollers for one paver should be expected.

- An extra roller with a side-roll for layer edges should be provided, especially with an increased number of construction joints or connections.
- Rollers fitted with a gritter for finishing SMA surfaces are also indispensable (see Section 10.7.1).

10.4.1.1.1   Vibratory Rollers

Vibratory rollers should be carefully or, more precisely, consciously used. An additional condition for using vibration is maintaining the appropriate temperature behind the paver to enable the movement of particles during rolling. Applying vibration to a cool SMA mixture is a mistake that leads to the crushing of particles. Vibration is not an option when SMA is being placed in a thin layer (i.e., 20–30 mm), on a stiff base (e.g., concrete slabs or a brick or block pavement), or at too cold of a mixture. Compacting SMA with vibratory steel rollers is generally permitted, but high frequency and low amplitude vibrations are a must (Asphalt Review, December 2004).

Determining the correct type of vibratory roller to use from among the following is important when considering the use of vibration:

- Classic vibratory rollers—used when it is certain that the frequency and amplitude of vibrations do not threaten to crush particles or to squeeze mastic out on the layer surface.
- Vibratory rollers with other vibration techniques
  - Oscillatory—marked by vibrations within a range of oscillations, compacting a shallower depth than classic vibratory rollers but sufficient for a wearing course
  - Variable direction of vibration—features the possibility of amplitude direction control.

The use of vibration can be an acceptable method of compacting SMA layers; most producers of rollers currently have solutions to minimize the risk of crushing an SMA skeleton. Obviously using vibration does not apply to the aforementioned situations (thin layer, stiff underlayer, too cold of a mix), *a priori* limiting the possibility of using vibratory rollers.

10.4.1.2   The Number of Rollers

The number of rollers, their types, and number of passes should be selected with regard to the SMA gradation, layer thickness, weather conditions, and the planned paving speed. Obviously the rollers should be in good working order, able to work at low and constant speeds.

The Minimum number of rollers to compact one layer one lane wide is two or three. Many nations' guidelines stipulate diverse requirements for the numbers of rollers, mainly stating that one should use rollers weighing more than 9–10 tons each. If three rollers are in use, two of them should be responsible for the majority of the compaction, and one should be the finisher. Extra rollers should be added in case of

TABLE 10.1
A Summary of the Usefulness of Equipment for SMA Placement

Operation	Steel Triple- Wheels Rollers	Pneu- matic Tire Rollers	Steel Tandem Rollers				Combi Rollers	
	Static	Static	Static	Vibration	Directional Vibration	Oscillation	Static	Vibration
SMA 0/11 mm and finer	Yes	No	Yes	Yes	Yes	Conditionally	No	No

Source: Based on M VA, Merkblatt für das Verdichten von Asphalt (M VA), Ausgabe, FGSV Köln, Germany, 2005.

problems with reaching the design layer density. Table 10.1 shows an overall summary of the usefulness of equipment for SMA placement based on German compaction guidelines M VA 2005.

10.4.2 Operating Rollers

The following sections outline some considerations regarding the operation of rollers once the roller combination for a specific job has been determined.

10.4.2.1 General Rules

In standard conditions, rollers should follow as closely as possible behind the paver. If it is not possible for the rollers to keep up with the paver, the speed of the paver should be reduced or the number of rollers increased. The method of rolling and positions of the different rollers are thoroughly discussed in the USACE Handbook 2000, *Dynapac Handbook,* and German DAV Handbook (Milster et al., 2004), where detailed directives can be found.

Rolling thin SMA layers should be executed with great caution, with vibration only rarely applied. The use of slightly lighter rollers instead of heavy ones is recommended.

The quantity of so-called SMA roll down (i.e., the change in thickness of a spread layer due to rolling) may be estimated as 10–15% of a layer thickness. It depends, among other things, on the design gradation curve that forms the aggregate skeleton.

10.4.2.2 Sequence

Various compaction sequences are adhered to. Generally speaking, every road-engineering company works out its own procedure after some time. By and large, the standard rule states that the paver is followed by static rollers first and then by vibratory rollers. Final passes are always carried out by static rollers, which finally level the surface, removing traces of rolling from it (the so-called finishing). When using

FIGURE 10.11 The particular way rollers approach the paver, with a turn being made right behind the paver. (Photo courtesy of Konrad Jabłoński.)

rollers with new types of vibration, it is worthwhile to consider the manufacturer's suggestions.

One of the more interesting techniques of rolling consists of the first roller behind the paver (the breakdown roller) making a slight turn when it approaches the paver (Figure 10.11). The remaining rollers operate without making the turn by simply reversing direction. This technique, among others applied in the *Kompaktasphalt*, is aimed at achieving better layer smoothness.

10.4.2.3 Speed of the Rollers

The speed of the rollers should be controlled and should be slow. According to the NAPA SMA handbook QIS 122, the speed of a roller during rolling must not exceed 5 km/h. In one of the British guidelines, the speed of the rollers should normally be between 4 and 6 km/h (SEHAUC, 2009).

Rolling with vibration at a reduced speed improves the effectiveness of compaction. However, doing so may lead to crushing particles and squeezing mastic out on the surface of a layer. Therefore great caution should be exercised, and vibration should be turned off if need be. Rollers with a speed control option enable automatic control to disengage vibrations during braking and changing direction. It should be kept in mind that rollers with speed control require a longer braking distance and skillful control.

10.4.2.4 Number of Roller Passes

Usually six to nine roller passes are enough to compact an SMA mixture. Moreover, compaction should not cause squeezing of the mastic onto the surface. The number of passes with vibrations should be limited to the indispensable minimum (most frequently, three).

On many construction sites or at the start of a new SMA mixture, it is recommended that the effectiveness of rolling in relation to the number of passes and type of rollers be tested on a specified section. The increase in density with more rolling passes is monitored with, for example, a nuclear density gauge, and the number of passes to reach the desired density is determined.

10.4.2.5 Time Available for Compaction

The time available for compacting the layer depends mainly on the following conditions during placement:

- The mixture temperature behind the paver
- The air and surface temperatures and wind velocity
- The layer thickness

The matters of temperature are elaborated on in Section 10.3.2.

In extremely adverse weather conditions, the time for compacting is counted in minutes and is often less than 5 minutes. After that time, the mixture temperature falls to a level at which the high viscosity of binder makes the movements of the mixture particles impossible. If the SMA layer has to be placed in difficult conditions, the following organizational issues should be remembered:

- Using the best possible insulation during SMA mixture delivery
- Closely coordinating the mixture deliveries in relation to the spreading speed with no stoppages and shutdowns of the paver or trucks delivering mixture to a work site
- Discharging consecutive mixture deliveries to the paver hopper before it is completely emptied
- Effectively heating of the paver screed
- Maintaining the proper paver speed to avoid the following:
  - Decreased effectiveness of the paver compaction
  - The risk of dragging chipping particles out by the screed
- Keeping the rollers close behind the paver and having more of them than when compacting in good conditions
- Calculating the limited time of effective rolling caused by the drop of mixture temperature in the layer; being aware of weather conditions, the temperature of a delivered mixture, and the layer thickness, one can roughly estimate the time required for compacting using ready-made curves, measuring the layer cooling rate, or calculating it with the use of computer software such as PaveCool or MultiCool.

10.4.2.6 Final Remarks

The layer edge should be rolled with a machine fitted with a side-roll; this will enable suitable compaction of the area close to the edge. The drums of a roller should be moistened with water, which should protect them against mastic adhesion and dragging particles out.

10.5 PLACEMENT OF SMA IN *KOMPAKTASPHALT* TECHNIQUE

The *Kompaktasphalt* method consists of placing two layers of a pavement in one pass of a specially designed paver (Figure 10.12). Typically the paver places both an SMA and asphalt concrete in one pass. The first attempt at such a laydown occurred

FIGURE 10.12 Special paver for the placement of two layers in one pass with an SMA wearing course (A2 highway in Poland), an example of *Kompaktasphalt* technology. (Photo courtesy of Konrad Jabłoński.)

in 1995 Germany and was made by Elk Richter of Fachhochschule Erfurt and the company Hermann Kirchner GmbH&Co KG on Highway A4 (Richter, 1997). In December 1998, also in Germany, a special modular paver was used for the first time (Utterodt and Egervari, 2009). This technology has been chiefly applied in Germany. Undoubtedly, the simultaneous placement of two layers has a number of strong points. It is enough to mention just the following:

- Excellent interlayer bonding (i.e., the "hot-on-hot" placement)
- Great thermal capacity of the two layers together (thickness of 10–12 cm), giving more time for compacting, especially on cool days
- Rapid progress in work (two layers at the same time)

The use of the *Kompaktasphalt* technology carries with it some additional requirements for the construction site organization:

- The mixture intended for a particular site comes from two, sometimes even three, asphalt plants with sufficient capacity to supply the paver.
- An adequate number of transport vehicles are essential to haul mixtures from asphalt plants to the construction site.

This frequently adopted scheme provides for the placement of an asphalt concrete intermediate layer and an SMA wearing course. Various thicknesses and gradation of SMA layers are used, with 4-cm of SMA 0/11 mm and 3-cm of SMA 0/8 mm being used most often.

10.6 TESTING THE FINISHED LAYER

A series of acceptance tests are carried out after finishing the placement of a mixture. They usually comprise measuring the content of air voids and the compaction factor, conducted on specimens of cores taken from the finished pavement. In many countries, nuclear density gauges are used for testing the homogeneity of compaction. When this is the case, cores are only required for calibration and comparisons with the nuclear gauge. Other properties checked on finishing the layer are skid resistance and the macrotexture depth. The methods used for these tests depend on national specifications.

10.6.1 Air Voids in a Compacted SMA Layer

The content of voids in the compacted SMA layer is the most commonly found parameter checked at the acceptance of a finished layer and is always mentioned as a fundamental. According to most analysis documents worldwide, the content of air voids should be lower than 6.0% (v/v). Lately in German guidelines ZTV Asphalt-StB 07 (September 2008 issue), this value has been lowered to 5.0% (v/v). This is the requirement most closely related to the durability of the compacted layer, including its susceptibility to water permeability, frost heave, and deicers. In countries with no significant drops of temperature below 0°C, higher contents of voids in the SMA layer (e.g., up to as much as 5–8% [v/v]) have usually been permitted.

An insufficient content of voids in a layer is also disadvantageous; recent experience shows that less than 3% (v/v) brings about the risk of premature rutting. It has been underlined in the literature (Voskuilen, 2000, Voskuilen et al., 2004) that when designing SMA with a determined content of voids in laboratory specimens (usually 3–4% [v/v]), one should remember that the final amount of voids in a compacted layer depends, among other things, on the arrangement of skeleton particles and voids among them. The air voids achieved in a layer on a work site are different from those achieved in the laboratory, just as particles compacted using a Marshall compactor are arranged differently than those that are rolled. Some authors (Voskuilen, 2000) also claim that during the life of the pavement, the content of voids in an SMA layer decreases due to the gradual decrease of voids in the chipping skeleton (e.g., postcompaction, crushing particles). That is why, for instance, in the Netherlands, designing SMA with an initial (laboratory) content of voids at the level of about 5% (v/v) has been practiced. Then, after some time of service, the air void content in the field was lowered to 2–3% (v/v) (Figure 10.13).

10.6.2 Compaction Factor

Almost all documents on SMA reviewed for the purposes of this book contain the compaction factor as a specified requirement. As in many other cases, the differences between European and U.S. specifications are clear.

That factor has been defined differently in different countries. The differences are grounded in the different reference density used related to the bulk density of the layer achieved on the construction site. That is the source of the sharp differences in numerical values: from 94% (the United States.) to 98% (Norway). The description of two basic definitions of the compaction factor follows later on.

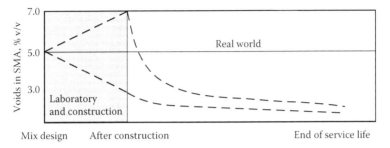

FIGURE 10.13   Change in the content of air voids in an SMA layer. (From Voskuilen, J.L.M., *Proceedings of the 6th International Conference Durable and Safe Road Pavements*, Kielce [Poland], 2000. With permission.)

10.6.2.1   Compaction Factor as a Quotient of Bulk Densities

In European practice, compaction factors based on a quotient of bulk densities have been commonly used. That quantity is calculated in percent according to the formula

$$c = \frac{\rho_s}{\rho_{sl}} \cdot 100\%$$

where
  c = compaction factor, %
  ρ_s = the bulk density of a specimen cut out of a pavement, g/cm^3
  ρ_{sl} = the bulk density of a specimen prepared of the same constituents and compacted in laboratory conditions (so-called reference specimen), g/cm^3

The most common requirement for the SMA compaction factor calculated according to this equation is c is at least 97%, though the requirement of c being no less than 98% can also been found.

A *sine qua non* for the correct calculation is taking into account the following assumptions:

* The SMA mixture used for preparing reference specimens compacted in a laboratory should come from the mixture being placed (e.g., from a truck just leaving the asphalt plant).
* Compaction conditions for reference specimens, specifically the temperature and compaction, should be adequate to meet the real conditions prevailing on the construction site.
* The control specimens cut out of the pavement should not come from areas close to the edges or at the beginning or end of a working lot.

The German guidelines for compacting asphalt surfacing (M VA 2005) have included additional procedures and directions to facilitate control of the compaction process for difficult-to-compact mixtures. The compaction factor of an asphalt layer, denoted as k, is calculated according to the equation:

$$k = \frac{\rho_s^b}{\rho_s^{lab}{}_{(E=50)}} \cdot 100\%$$

where

ρ_s^b = the bulk density of a specimen cut out of a placed layer

$\rho_s^{lab}{}_{(E=50)}$ = the bulk density of a specimen prepared in a laboratory with the use of 50 strokes of a Marshall hammer on each side of a specimen

The bulk density of laboratory specimens is determined on material prepared with 2×50 strokes of a Marshall hammer. The compaction temperature should be appropriate for the applied binder. Due to various interpretation problems and mistakes in the calculation of the index k, the new German guidelines (M VA 2005) have introduced an additional parameter—the compaction index K referring to the bulk density achieved on the construction site relative to the maximum density

$$K = \frac{\rho_s^b}{\rho^\infty} \cdot 100\%$$

where

K = compaction index, %

ρ_s^b = the bulk density of a specimen cut out of an executed layer

ρ^∞ = the maximum analytical density of a given mixture

Having substituted the maximum density, the values of the compaction index will be higher than 94%. The method of using the quotient of two bulk densities for determining the SMA compaction factor has been called out by the European standard EN 13108-20. Its clause C.3 stipulates that the reference density for indicating the compaction factor is the bulk density. The standard also provides that detailed conditions for preparing specimens and determining that density using EN 12697-5 and EN 12697-6 shall be declared.

10.6.2.2 Compaction Factor as a Quotient of the Bulk and Maximum Densities

Calculating the quotient of the bulk and maximum (the so-called theoretical maximum density [TMD] or the Rice density after ASTM 2041) densities is a popular method of determining the compaction factor. The requirement for compaction of an SMA layer is a minimum 94% (of the maximum density) according to the NAPA SMA Guidelines QIS 122; that makes upto 6% (v/v) of air voids allowable after compaction. In this case the reference density does not depend on conditions for preparing laboratory specimens. As in the previous method, densities obtained in a laboratory during mix design are not taken into account; results from testing the density of a mixture taken from an asphalt plant during trial production are recorded.

The essential strength of that method is making the compaction factor free from various conditions for preparing specimens. The possibility of making mistakes in determining the reference density in a laboratory obviously disappears.

10.6.3 MACROTEXTURE

In some countries, particularly where surface properties are high priorities for a wearing course, in addition to other parameters, requirements for macrotexture are also laid down. The British requirements (HA MCHW, 2008) for mixtures with an upper (D) aggregate size of 14 mm or less, stipulate a minimum 1.3-mm macrotexture depth[*] (measured using the volumetric patch method described in EN 13036-1) for high speed roads at the moment of opening the road to traffic. In some other countries the requirement for the macrotexture depth is a minimum of 1.0–1.2 mm, but this requirement usually does not apply to fine graded SMA 0/7 or 0/8 mm.

10.6.4 NUCLEAR GAUGE DENSITY MEASUREMENTS

In many countries, nuclear gauges are used for field testing compacted asphalt layers. If well-calibrated with core samples, nuclear gauges are convenient tools for rapid testing of field density.

In a report by Brown and Cooley (1999), they pointed out that the application of leveling sand and dynamic correction factors improve the accuracy of the density tests.

British guidelines (BS 594987:2007) provide a protocol for calibrating and operating indirect density gauges, including gauge operations, initial calibrations, and consistency of calibration.

10.7 FINISHING THE LAYER

Some countries require special finishing techniques for SMA surfaces. Those requirements are described here.

10.7.1 GRITTING

Applying a layer of grit to the finished surface is often specified. Why has gritting been applied? Soon after placement, the SMA layer is characterized by relatively low friction (the so-called postconstruction slipperiness) caused by a thick film of binder on particles of aggregate. Spreading additional aggregate on the surface of the hot SMA, followed by rolling (to embed the grits) is aimed at breaking the binder film on the coarse particles. Due to its microtexture, well-embedded grit breaks the water film, hence increases the skid resistance of SMA.

Without gritting, the process of rubbing the binder film off the aggregate particles that provide the SMA macrotexture proceeds slowly under the action of traffic, first in the wheel paths, and then, after some time, all over the roadway surface. This pattern of wear brings about the development of nonuniform friction characteristics of

---

[*] Average per 1,000 m section: not less than 1.3 mm; average for a set of 10 measurements: not less than 1.0 mm (HA MCHW, 2008).

the wearing course. Dutch researchers (Jacobs and Fafie, 2004) have demonstrated low values of friction coefficients appearing on both a dry and a wet SMA course:

- On a dry pavement in summer, the binder film on the surface of the aggregate particles softens at high temperature, changing the binder film into a lubricant, reducing the friction between tires and aggregate particles.
- On a wet pavement, because of the thick binder coat, the aggregate microtexture is not able to break a water film on the surface of an SMA course. If the SMA contains a PMB, the binder film remains intact for a longer period of time.

Therefore some additional solutions for enhancing the friction coefficient should be applied to increase the friction in the early stages of trafficking the roadway while the binder film remains intact. Some observations prove that it might take as long as 6 months to rub off the binder film on the coarse particles.

10.7.1.1 Gritting Materials

Grits usually consist of different types of aggregates, such as the following:

- Aggregate 2/5 (or 2/4) mm, cleaned and hot, applied at a rate of 1.0–2.0 kg/m², for SMA with a gradation of at least 0/11 mm (Figure 10.14)
- Aggregate 1/3 mm, washed and hot, applied at a rate of 0.5–1.5 kg/m², for SMA with any gradation
- Aggregate 0.25/2 mm, washed and hot, coated with about 1% binder (m/m), applied at a rate of 0.5–1.5 kg/m², for SMA with any gradation but preferred for SMA 0/5 and 0/8

Generally, the use of finer grains for gritting means a lower quantity per square meter. Additionally the smaller the maximum aggregate size of the SMA, the smaller the grit grains that should be used.

FIGURE 10.14 Gritting particles among SMA coarse aggregates. (Photo courtesy of Krzysztof Błażejowski.)

Gritting with aggregate 2/5 (2/4) mm is not recommended for SMA in Germany because it brings about an increase in noise generated by the contact between tires and the gritted pavement, (Drüschner and Schäfer, 2000). However, Dutch research has not proved an increase in noise caused by the use of grit (Jacobs and Fafie, 2004).

Gritting with an aggregate coated with binder provides a durable bond between the grit and the hot SMA layer. Attention should be paid to preventing an overdose of grit to avoid its sticking to the drums of a roller and destroying the hot layer surface. An overdosing of grit uncoated with binder can produce a similar effect (Jacobs and Fafie, 2004).

It is worth knowing that in British guidance (HAUC 2009), the aggregates for gritting should have a PSV not less than 55, which is more than the PSV of coarse aggregates used in an SMA skeleton in many countries (e.g., Germany, Poland, Hungary).

10.7.1.2   Gritting Execution

Grit is spread on an hot SMA mixture using one of the two following techniques:

- With a gritter installed on a roller, during the first pass of the roller there is no gritting, but during the second pass the gritting is turned on. One should remember to grit in one direction of roller movement, and to not grit when returning; when using that technique, grit particles are pressed into the hot SMA mixture (Figure 10.15).
- With a self-propelled gritter, gritting starts after rolling the layer down with the rollers, when the SMA is still hot enough and prior to the last pass of the roller. This method is seldom used.

Remember that the quantity of grit per square meter has been established for a given machine (a gritter) and for a specific passing speed. Any change to the speed

FIGURE 10.15   SMA gritters installed on rollers—the gap gritter. (Photo courtesy of Bartosz Wojczakowski.)

FIGURE 10.16   Movement traces of a roller after gritting a freshly made SMA layer. (Photo courtesy of Krzysztof Błażejowski.)

results in a change in the quantity of grit applied to the layer. The minimum temperature of a layer being gritted is 110°C; below that temperature the grit is unlikely to stick to the SMA (Jacobs and Fafie, 2004).

Figure 10.16 shows movement traces of a roller fitted with a gritter. Spots where the roller has changed direction before the gritter was fully disengaged are visible. This is not a substantial mistake of execution; at most it is likely to spoil the aesthetic impressions of some road users. After gritting and cooling of the layer, excess grit aggregate or unbonded grit should be removed, (e.g., by sweeping).

Research has proved that gritting makes sense since it enhances SMA antiskidding performance in the initial period after pavement construction. When gritting is not used, better coefficients of friction are available with SMA 0/8 mm than with SMA 0/11 mm (more contact points between SMA and tire). It should be mentioned additionally that the application of gritting influences the macrotexture during the first period after construction. Grains and crushed particles of grit occupy spaces between coarse grains and decrease the macrotexture depth.

Finally, it should be clearly stated that gritting is not a way to cover or hide fat spots occurring on SMA surfaces during placement. In some cases, even gritting is not likely to conceal a fat spot, since gritting would be "drowned" in an excess of mastic (Figure 10.17).

FIGURE 10.17 Gritting of chippings 2/5 mm applied over a fat spot of mastic. (Photo courtesy of Krzysztof Błażejowski.)

10.7.2 Edge Sealing

In regard to edge sealing, the following reasonable solution definitely extends the durability of a pavement and has been adopted in German practice (Milster et al., 2004):

- An extra layer of tack coat (of hot binder) is placed on the edges of pavement (between the layers), minimum 10 cm wide.
- A binder layer, minimum 2 mm thick, on a side surface of all layers, is applied (to achieve such a layer of binder, the tack coating usually needs to be repeated, or a hot sealant 2 mm thick needs to be applied in one working cycle).

This sealing is aimed at protecting the pavement against water infiltration between layers and inside them.

10.8 OPENING TO TRAFFIC

Before opening up a road section with a new SMA layer to traffic, one should be sure that it has sufficiently cooled. Letting traffic go on an SMA layer while it is still warm may cause its premature rutting and the squeezing-out of the mastic. Some regulations stipulate that the opening to traffic may happen no sooner than 24 hours after finishing the placement, or at least, when the temperature in the middle of a layer drops to 30°C. The common practice is to delay the opening to traffic until the moment that the layer temperature drops to the air temperature. When an earlier opening to traffic is necessary, only light vehicle traffic should be allowed.

11 Problems

As with other asphalt mixtures, problems frequently occur with SMA, too. They may develop at the stage of design, production, or application. Some of the troubles described here also appear in asphaltic concrete or other asphalt mixtures, but a distinctive feature of SMA is that its shortcomings are revealed in a particularly clear, and sometimes painful, way.

This chapter presents an arranged collection of the most common defects observed in SMA. But before we begin, two issues need to be made clear:

- Each group of problems and their probable causes are provided here based on the author's subjective assessment derived from his experience and consideration. Therefore it is obvious that a reader may have a different view on a given problem or that it may have another cause.
- Always and in each case, while examining a detected problem, one should be open to investigating its various causes, including the less obvious ones. Nobody should limit himself or herself to the pieces of advice contained in this book; an open mind is the recommended approach.

An immediate "blessing" when a problem occurs is having a laboratory or possibly the asphalt plant staff available to help correct the problem; these resources are becoming more widely available. The lab or plant staff can help to evaluate the materials and operations to identify the causes of the problems, such as changes in material properties, plant temperatures, moisture conditions, etc. This chapter will make it apparent that the members of the team preparing road machines and the site team are as important in the final outcome (success or failure) as other elements.[*]

Now it is time to examine with the problems in detail.

11.1 LONGITUDINAL FAT SPOTS OF BINDER

Longitudinal fat spots are the most frequently seen defects when using SMA. Such spots may be defined as areas with an excess of binder or mastic that are shaped longitudinally and are parallel to the path of the paver.

There are two types of longitudinal fat spots—binder fat spots and mastic fat spots—which differ only in the content of the fat spot's components.

Longitudinal fat spots (Figure 11.1) contain, firstly, some amount of binder appearing on the surface of a course. They are easily recognizable simply by scratching the fat spot with the metal stem of a thermometer. A thin binder layer

---

[*] To illustrate some problems, infrared images were used (with temperatures given in degrees Fahrenheit). However, not all of them concern SMA layers; the goal is to show the temperature differences during laydown.

FIGURE 11.1   Longitudinal fat spots of binder. (Photo courtesy of Bartosz Wojczakowski.)

is visible on the surface of the SMA course, with regular SMA underneath. Thus such fat spots are similar in appearance to mastic fat spots but with no separation of all the mastic from the coarse aggregates. They arise because of an excess of unbonded binder. They may be caused by an error during one (or more) of the following:

- Design (too much binder designed in SMA)—this type of error should be detected by the design laboratory during draindown testing; thus it happens relatively rarely.
- Production—it is more likely that an overdosage of binder will occur during mixture production (e.g., due to an inaccurate weigh scale for the binder) or because the metered amount of stabilizer proved to be too small to prevent binder draindown—that is, an inaccurate stabilizer metering system or inadequate action from the stabilizer (e.g., a granulated stabilizer of poor quality or one damaged by an excessive time of dry mixing*).
- Change in the properties of the components—this can occur any time a component is changed during the course of SMA manufacture.
  - Fillers are changed and have significantly different contents of voids than previous fillers.
  - PMB viscosity is too low at the production and laying temperatures (occurs especially when PMB is used for an SMA without fibers).
  - Granulated stabilizer quality changes or is overpressed.
- Surface preparation—in some circumstances an excess of binder coming from an underlying tack coat (which may result from excess tack coat binder gathering in depressions) or mastic asphalt patches situated under SMA may be drawn up into the SMA surface by the heat of the newly placed course.

---

* It should be pointed out that an excessive dry mixing time of loose stabilizers causes abrasion of fibers, reducing them to filler (dust) and substantially lowering the stabilizer's binder absorption effectiveness.

11.2   LONGITUDINAL FAT SPOTS OF MASTIC (SEGREGATION)

Mastic fat spots may be distinguished from binder fat spots using a tool to check the composition of a particular fat spot. All the elements that make mastic—binder, filler, fibers and sand—can usually be identified in a cross section of a fat spot. It is necessary to examine what is going on around the fat spot to discover its cause. Usually there is one of two sets of conditions—with segregation or without segregation.

11.2.1   Fat Spots of Mastic with Segregation

An area with a definite shortage of mastic adjacent to the fat spot (visibly porous) is a clear sign that this is a case of an SMA segregation—namely, separating the coarse aggregates from the mastic. If the quantities of all the ingredients have been properly selected, the high accumulation of one component in one location will result in a reduced quantity of that component in another. When an excess of mastic (fat spot) appears somewhere, the coarse aggregate content rises elsewhere, so the total sum of the components remains constant. Figure 11.2 presents a classic fat spot in a segregated SMA mixture. Figure 11.3 shows the difference between fat spot and adjacent porous section and a close-up of the mastic-rich area.

It is worth emphasizing that such segregation may happen for one or more of the following reasons:

• Substandard production of an SMA mixture
  • Lack of a stabilizer, its improper metering, or its poor quality
  • Excessive production temperature
  • Too short a time of mixing components
• Too long a storage time of a mixture in an asphalt plant storage silo
• Improper laydown of SMA (with an improper setup of a paver)

FIGURE 11.2   SMA mixture segregation—separation of mastic from coarse aggregates. (Photo courtesy of Krzysztof Błażejowski.)

(a)   (b)

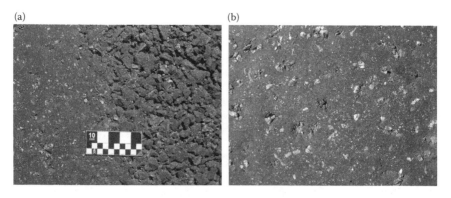

FIGURE 11.3   SMA mixture segregation—separation of coarse aggregates from mastic: (a) a junction of porous and closed parts enlarged and (b) surface of separated mastic enlarged. (Photos courtesy of by Krzysztof Błażejowski.)

11.2.2   FAT SPOTS OF MASTIC WITHOUT SEGREGATION

If porous spots are not visible around the fat spot, this is not a case of classic segregation. We may suppose that the fat spot appeared because of an excess of mastic volume (an error of design). It may also have been caused by too much energy applied during compaction (improperly adjusted or unskillfully applied vibration); this will lead to the crushing of the coarse aggregate skeleton and the squeezing out of the mastic. Then the surface with the bleeding mastic will make up a major part of the surface (which, simply put, is any area where bad compaction was applied). A related problem, described in Section 11.5, describes an SMA surface that is too closed or tight.

11.3   SMALL CIRCULAR FAT SPOTS

Small fat spots with a circular shape of approximately 3–20 cm in diameter may appear on a laid SMA course, randomly scattered without any visible order of occurrence (Figures 11.4 and 11.5). This type of fat spot hardly ever appears. As usual, their composition should be checked. Most often, mastic and stabilizer (especially loose fibers) can be found in them. They are most often caused by insufficient mixing of the SMA components (i.e., insufficient wet-mixing time at the mixing plant).

11.4   STABILIZER LUMPS

Stabilizer lumps occur rarely, but sometimes they bring fairly unpleasant consequences. The problem lies in the fact that symptoms may only be seen some time after the placement of an SMA course. Results such as the following are quite spectacular:

* Initial formation of bulging spots in a newly made course
* A spot of mixture that separates from the mat

Both cases are caused by lumps or pieces (Figure 11.6) of an undistributed (unmixed) stabilizer of loose fibers. Moisture absorbed by a stabilizer is able to lift

FIGURE 11.4 An example of randomly scattered circular fat spots of about 1–15 cm in diameter. (Photo courtesy of Krzysztof Błażejowski.)

FIGURE 11.5 A circular fat spot about 20 cm in diameter, enlarged. (Photo courtesy of Krzysztof Błażejowski.)

FIGURE 11.6 A big lump of stabilizer incorporated into an SMA course, after removal. (Photo courtesy of Krzysztof Błażejowski.)

FIGURE 11.7   Stabilizer lumps incorporated into an SMA course. (Photo courtesy of Marco Schünemann.)

FIGURE 11.8   The closed structure of an SMA course over a large area. (Photo courtesy of Krzysztof Błażejowski.)

an SMA course enough to be discernible to the naked eye. It also causes the mixture to ravel and create pot holes (Figure 11.7). These situations may happen as a result of the following:

- Wet, loose fibers were applied.
- Loose fibers was added into the pugmill too late—namely, after binder dosing.
- Wet-mixing time was excessively shortened.
- Stabilizers of poor quality were used.

11.5   TOO CLOSED SMA STRUCTURE

It happens that, after the completion of an SMA course, a sight reminiscent of mastic asphalt may unfold before the observer's eyes (Figure 11.8). Such a closed SMA structure means utter failure and the likely removal and replacement (possibly by recycling) of the layer. This kind of problem is directly connected to the discussion in Section 11.2 about fat spots of mastic (without segregation) but usually occurs over most or all of the layer.

Reasons for such a significant setback include the following:

- Errors of mixture design
  - Too much mastic in relation to the voids' volume among the compacted coarse aggregates
  - Upward designing of a gradation curve in the filler–sand fraction
  - Use of weak aggregates
- Errors of compaction—application of excessive compaction energy (too high a vibration amplitude), which causes the crushing of grains and their closer arrangement, thus reducing voids meant for mastic
- A combination of these two causes—an application of weak aggregates is conducive to such a superposition; then we can speak about an error of design (selection of aggregates) and an error of placement (excessive energy of compaction).

If we happen to observe a large, closed SMA surface accompanied by a distinctive odor, it is worthwhile reading Section 11.11.

11.6   TOO POROUS SMA STRUCTURE

Just as it is possible to find an SMA course that is too closed, so too it is possible to find a course that has too high a void content. This problem may occur over very large areas, which are marked by excessive porosity. Local porosity over smaller areas, is described in Sections 11.8.2.3 through 11.8.2.9.

The issue of how open the SMA structure should be has been debated for some time. True enough, we happen to find a quite porous SMA structure every now and then. An investor or may owner sometimes agrees to leave in place an SMA that is too closed (after an antiskid treatment); unfortunately an SMA that is too open more often falls victim to a road-milling machine. An open structure (Figure 11.9) of a mixture consisting of lots of mutually connected pores is conducive to water and air penetration (see Chapter 12), which results in a shorter service life of the course.

An excessively open structure of an SMA course may be caused by any of the following:

- An error in designing the mixture (e.g., high coarse aggregate content, insufficient volume of mastic)
- An error while mixing at the asphalt mixing plant (e.g., incorrect composition of the mixture [i.e., not in conformity with the design])
- An error of placement (e.g., paver setups, manual spreading of the mixture, insufficient temperature of the mixture during laying, undercompaction)

FIGURE 11.9   A close-up of an SMA course surface with a high content of voids. (Photo courtesy of Krzysztof Błażejowski.)

High contents of free voids in a compacted SMA course may result from a mistake made while designing the mixture. Usually the cause is a result of one of the following:

- An undue shifting of the grain size curve to the right and an increase of the coarse aggregate fraction to greater than 80%, with a simultaneous decrease in both the sand fraction and the filler, plus a reduced content of binder—as we remember, an increased amount of material on the 2.0 mm sieve requires an increase in the binder content to fill the voids of an SMA mixture
- A reduction of SMA binder contents, sometimes practiced for the sake of economy
- The application of excessive temperatures and increased energy during the compaction of samples in the laboratory,* resulting in an incorrect mixture design.

If porosity appears over larger areas and is not an error of composition, one may expect that the number of rollers or passes have not been suitably selected or that too cool a mixture has been spread. Sometimes the difference may be noticed when watching the layout structure of aggregate grains. Grains of a mixture spread at too low a temperature look as if they have been pulled by the paver screed; they are not arranged tightly side by side (Figure 11.10).

11.7   SMA SEEPAGE AND PERMEABILITY TO WATER

SMA surface seepage is directly related to excessive porosity of a compacted mixture (see Chapter 12) and the condition (watertightness) of the course situated just under the SMA. When assessing the watertightness of an SMA course, one must not forget that water permeability is determined not only by the contents of voids on the surface of a compacted course but also by the shape and interconnectedness of the inner pores. Finally, most national regulations and numerous publications recommend a

---

* That aspect of problems in design is discussed in Chapter 8.

FIGURE 11.10 The surface of an SMA course just after spreading at too low a temperature. (Photo courtesy of Krzysztof Błażejowski.)

FIGURE 11.11 Seepage of water through an SMA layer. (Photo courtesy of Bohdan Dołżycki.)

limit of 6.0% (v/v) air voids in a compacted course, above which a course becomes partially permeable to water.

A tight intermediate course under the SMA, which prevents the penetration of water and water vapor deep into the pavement, is conducive to the development of SMA surface seepage (Figure 11.11).

11.8 PROBLEMS RELATED TO TEMPERATURE AND LAYING TECHNIQUES

Thermal problems occurring during the manufacture and placement of SMA are often underestimated, and yet they can lead to significant defects in a new course and reduce its working lifetime. It has already been indicated in Section 11.6 that the placement of a cool SMA mixture may result in an excessive content of voids.

Aside from problems directly connected with the temperature of a mixture, the effects of wrongly adjusted equipment during placement may be observed. Now and again errors in laying and the effects of an unsuitable mixture temperature superimpose onto each other. These problems may be divided as follows:

- Manufacture and transport
  - Unsuitable temperature of an SMA ingredients during manufacturing and storing in a silo
  - Wrong methods of transporting a mixture
- Laying
  - Improper temperature of the supplied mixture
  - Adjustment errors by the paver and in general methods of laying

Infrared cameras are increasingly being used to analyze thermal problems. Results of such research efforts may be found in several publications (Pierce et al., 2002; Stroup-Gardiner et al., 2000; Willoughby et al., 2001).

11.8.1 PROBLEMS OF MANUFACTURE AND TRANSPORTATION

Most questions related to maintaining a suitable temperature during the mixture's manufacture have been regulated by commonly known technical specifications (e.g., standards, guidelines). These questions are discussed in Chapter 9. Simply speaking, there are two main problematic instances in the matter of SMA manufacture temperatures—when the temperatures are too low or when the temperatures are too high.

Too low a manufacturing temperature of a mix prevents the formation of a correct and complete asphalt binder film. Too low a temperature can be visually recognized; aside from visible uncoated aggregate pieces, the appearance of the mixture seems to be matte. Since SMA contains a considerable amount of binder, it should be glossy and glisten "glow wormly" (namely, like glow worms) at a suitable temperature.*

Too high a manufacturing temperature increases the threat of binder draindown from the aggregate and excessive aging of the binder. Loading the overheated SMA into a silo runs the risk of getting into trouble with excessive draindown. To check SMA susceptibility to draindown after mixture overheating, one ought to test it when the SMA is at a temperature 15°C (or 25°C) higher than the recommended manufacture temperature (see Chapter 8).

11.8.1.1 Mixture Production in an Asphalt Plant

When storing a finished, hot mixture in a silo, the mixture closest to the walls of the container is subject to gradual cooling. As storage time increases, the amount of cool mixture is likely to rise substantially. If the SMA contains modified binder, its

---

* This remark concerning mixture production and look applies to classic hot mix technology, not the so-called warm mix technology.

stiffening process proceeds much faster. Moreover, we may face additional trouble with cool mixture stuck to the walls of the silo unless the silo is heated. Also, if the silo chutes are not heated, problems with discharging the mixture out of the silo can occur.

11.8.1.2 Transport of a Mixture to the Working Site

When it is cold and the construction site are far away, we usually console ourselves with the idea that the mixture is hot inside so the cool crust may be somehow stirred into the hot mix and warmed up in the paver. The mix really is hot inside, as shown in Figure 11.12. Typical cracks in the cool crust formed on the surface of the transported mixture and its deeper cooling down by the truck's sideboards may also be noticed. That is why it is always worth using trucks that have well-insulated boards and are tightly covered.

In cold seasons there are little chance that delivered mix will have thermal uniformity. Usually after discharge of the mixture into the paver hopper there are many cool or poorly heated fragments in the mixture. Frankly, it is not wise to be under the illusion that the cool pieces can be stirred into the mix and warmed up and that the mixture will again be homogeneously hot (though the use of MTV vehicles can help—see Chapter 10).

11.8.2 Construction-Related Problems

The laying of too cool a mixture may result in a course with a considerably large amount of air voids (see Section 11.6). That is an obvious matter; however, quite often the accelerated cooling of a mixture takes place in particular areas during its laying. Such places have been defined as having porosity of different types.

The figures in the sections that follow are photos taken with an infrared camera. The Fahrenheit scale of temperature has been placed on each of them. They show various causes of mixture cooling, followed by thermal differences resulting in poor quality.

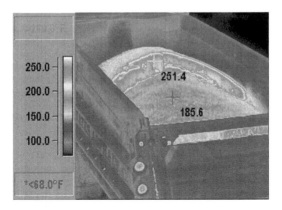

FIGURE 11.12   Distribution of asphalt mixture temperature after transport to the work site during discharge into the paver. (Photo courtesy of Kim A. Willoughby, WSDOT, United States.)

11.8.2.1   Paver-Made Streaks

Let us suppose that an adequately hot mixture is delivered to the job site. Nevertheless, badly adjusted paver elements (in the given examples, badly arranged screed segments) may bring about the formation of porous areas in the course (streaks). Figure 11.13 presents an example of streaks pulled by a paver during laying. Figure 11.14 presents the same effect but photographed with an infrared camera.

11.8.2.2   Paver Standstills

One of the main principles of laying SMA is avoiding a paver standstill. Almost every time the paver is immobile, uncompacted areas behind the machine are the result (Figure 11.15). This effect is heightened if the mixture contains modified binder (higher stiffening of the mixture during cooling down). Figure 11.16a and b presents infrared camera photos of mixture cooling down during a paver standstill.

11.8.2.3   Continuous Porosity by the Edge

This defect appears in a continuous way along the edge of a layer (Figures 11.17 and 11.18). It is mainly brought about by the following:

FIGURE 11.13   Formation of streaks or mixture pulling due to incorrect setup of paver screed segments.

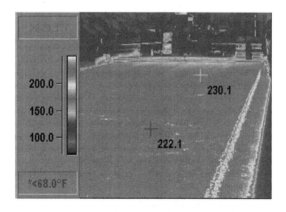

FIGURE 11.14   The effects of an incorrect paver setup, temperature in degrees Fahrenheit. (Photo courtesy of Kim A. Willoughby, WSDOT, United States.)

FIGURE 11.15   Area of an uncompacted cool mixture during a paver standstill (a zone of cooling mixture that is inaccessible to rollers).

(a)   (b)

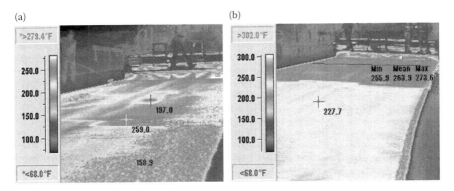

FIGURE 11.16 Paver standstills and their effects—infrared images. (Photos courtesy of Kim A. Willoughby, WSDOT, United States.)

FIGURE 11.17 Continuous porosity along the edge of a paved lane.

• Badly adjusted mechanism of mixture distribution through the augers or improper operation of the augers (insufficient quantity of a mixture supplied to the screed edge)
• Length of augers too short in comparison with the board width (several board segments with no additional feeding segments applied)

The porosity of a layer is usually easily seen when the surface is wet (Figure 11.18.b). Another problem (less often observed) can be caused by manual raking of a mixture along the edges of a spread lane (Figures 11.19 and 11.20). Unfortunately, leveling a newly spread mixture with rakes or shovels is the customary manner of some paver teams. Manual raking cannot produce the same level of uniformity as mechanical leveling, leading to cooler portions and porous areas in the mat.

11.8.2.4 Longitudinal Porosity behind the Center of the Paver
Figure 11.21 schematically represents a phenomenon caused by the improper distribution of a mixture with augers. The distribution of a mixture along a paver screed should be checked. If the mixture is not properly distributed across the full width of the paver screed, streaks of porosity may become obvious. Another cause may be the location of the augers' driving unit (gear box) in the middle of its width.

11.8.2.5 Longitudinal Porosity (off Center)
Longitudinal porosity parallel to the direction of spreading but not centered behind the paver appears fairly often (Figures 11.22 through 11.24). It may be caused by either of the following:

(a)   (b)

FIGURE 11.18 Continuous porosity along the edge of a spread mixture: (a) an infrared image and (b) porosity along the edges of layers—pockets of water visible during pavement drying after rainfall. (Photo [a] courtesy of Kim A. Willoughby, WSDOT, United States; [b] courtesy of Krzysztof Błażejowski.)

FIGURE 11.19 Manually corrected edges of a working lane with cooler temperatures and higher porosity.

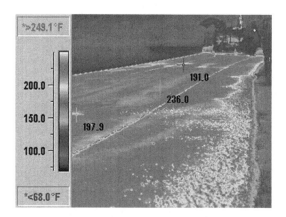

FIGURE 11.20 Position of manually corrected and leveled edges—an infrared image. (Photo courtesy of Kim A. Willoughby, WSDOT, United States.)

FIGURE 11.21 Longitudinal porosity behind the center of the paver.

FIGURE 11.22   Longitudinal porosity off the center of the paver.

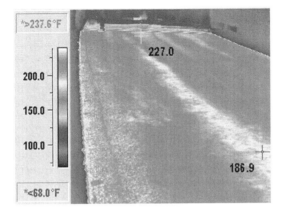

FIGURE 11.23   Examples of drop in mixture temperature due to longitudinal porosity in the spread lane. (Photo courtesy of Kim A. Willoughby, WSDOT, United States.)

(a)   (b)

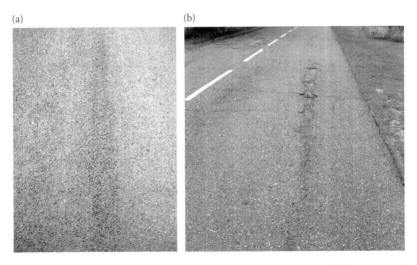

FIGURE 11.24   Longitudinal porosity: (a) on a new SMA pavement and (b) pavement damage in a porous area. (Photos courtesy of Krzysztof Błażejowski.)

- The augers distributing the mixture in a paver are positioned incorrectly (see Chapter 10), (i.e., unevenly along its screed). A solution to this problem could be to change of the setup of the augers.
- Pieces of various sizes of cool mixture get under the paver screed and pull the mixture, sometimes for quite a distance.

11.8.2.6 Porosity in Other Distinctive Places

Porosity may also be found as follows on an SMA course:

* At the beginning or end of a work site
* In irregularly shaped spots
* At the edge of a compacted course

Porous spots (areas) occur fairly frequently at the beginning or end of a work site when spreading an SMA (or other asphalt mixture) course. This chiefly results from some difficulties with the correct start of laying by a paver (sometimes called take off). Because the beginning of a work site cannot be cut off and removed, such areas of porosity appear mostly at the beginning of a spreading shift (Figure 11.25).

A similar result happens with the manual spreading of mixtures for some length at the beginning of paving followed by compacting with rollers. Manual work cannot equal the efficiency of placement with a paver, which provides more than 80% of the initial compaction. The effect is evident—the course in that area is uncompacted and uneven. Other places that were manually placed because there was no room for a paver look similar (Figure 11.26).

11.8.2.7 Spot Porosity

As mentioned before, pockets of porosity appear cyclically now and then (Figures 11.27 and 11.28). Their cause is not easy to explain. Maybe the paver was setup incorrectly or, more frequently, perhaps there were pieces of cool mixture in the SMA or segregation was still occurring in that stage of mixture production. Or the cause may be something entirely different and difficult to determine.

11.8.2.8 Transverse Porosity

Transverse (or lateral) porosity can often be seen on pavements made during unfavorable weather conditions (in autumn or in winter) or when the mixture has been

FIGURE 11.25 Beginning of a work site—effect shown years later, a consequence of spreading too cool a mixture at the beginning of the work site. (Photo courtesy of Krzysztof Błażejowski.)

FIGURE 11.26   Manual spreading of the SMA mixture on a down ramp from a side street. (Photo courtesy of Krzysztof Błażejowski.)

FIGURE 11.27   A diagram of spot porosity.

(a)   (b)

FIGURE 11.28   Local porosities: (a) an infrared image; (b) the effects after a couple of years on another road that has the same problem, arrows point at spots of pavement pot holes occurring at cyclic distances. (Photos [a] courtesy of Kim A. Willoughby, WSDOT, United States; [b] courtesy of Krzysztof Błażejowski.)

transported from far away. It has the distinctive appearance of transverse strips with increased porosity (Figure 11.29).

The most common cause of this is the buildup of cool mixture at the wings of the paver hopper. The distance between the strips of porosity in Figures 11.30 and 11.31 more or less reflects the distances between the loading places of fresh mixture into the paver from end dump trucks. This develops when the paver crew lifts the

FIGURE 11.29   A diagram of transverse porosity.

FIGURE 11.30   Infrared images showing porosity in the form of transverse, cyclic strips, temperature in degrees Fahrenheit. (Photo courtesy of Kim A. Willoughby, WSDOT, United States.)

FIGURE 11.31   An example of cyclic strips of transverse porosity on a new SMA course. (Photo courtesy of Krzysztof Błażejowski.)

hopper's wings to use all of the previously delivered material before accepting a new delivery. Since the older mixture is significantly lower in temperature, after it passes through the paver this cool mixture appears as a transverse streak of porous (cold) material behind the screed.

Various preventive methods may be used to avoid this porosity, including the following:

- Paver staff should regularly remove the mixture residue on hopper wings or leave that mixture in the hopper wings until the end of the day, when it can be removed and disposed of.
- Complete unloading should not be allowed; the fresh, hot mixture should be unloaded from the truck to the hopper while some of the previous delivery of mixture is left in the hopper.

11.8.2.9   Cracking of SMA Course When Rolling

Cracking of the mixture under rollers may be observed when there is no bond between an SMA course and the layer under it or when the temperature of the SMA mixture is too high. The temperature of the mixture during compaction is easy to control, therefore it will not be a topic of this section.

Figure 11.32 shows an old cement concrete pavement without a tack coat covered with a new SMA overlay. This resulted in a "dry slide" and the tearing of the SMA.

FIGURE 11.32   An example of the tearing of an SMA mat due to the lack of a tack coat on the cement concrete bottom layer. (Photo courtesy of Krzysztof Błażejowski.)

A similar accident might happen while paving SMA on an old asphalt course with a very polished surface. By contrast, a course with too much tack coat may lead to a "wet slide" on a layer, also resulting in the tearing of the mat.

The conclusion to be drawn from these examples are self-evident—SMA paving should be preceded with proper tack coat. The amount of tack coat should be carefully selected, taking into consideration the state of the underlying layer. Examples are shown in Chapter 10.

11.8.2.10 Unevenness or Irregularities

The last problem encountered when spreading SMA (and other asphalt mixtures, too) is unevenness caused by the approach of a paver on mixture residue left over on a bottom layer. Figure 11.33. shows how such residue builds up.

Different type of scrapers, sweepers, and other similar inventions fixed to the paver cannot fully protect it from causing unevenness. The care of cleanliness of the bottom layer is the responsibility of the paving crew. It consists not only in the skillful handling of a shovel but the proper coordination of mixture delivery from the dump truck into the paver hopper.

The material that is dumped in front of the paver also can cool before compaction, causing an internal porous area that can later trap water and then disintegrate the SMA layer.

11.8.2.11 Summary

Although thermal problems appear on almost every construction site and apply to any type of mixture, they seem to be underestimated. Examples of pavement damage presented in the previous pages, which developed during the construction stages, should make us aware of their power to reduce a new pavement's durability.

The use of infrared cameras has its strengths because it enables the spotting of potentially weak areas during construction when corrective measures can still be

FIGURE 11.33 Remains of mixture left on a bottom layer resulting in unevenness of the finished course. (Photo courtesy of Krzysztof Błażejowski.)

taken. Finally, it is worth finding the causes of damaged pavements that have pot-holes, bumps, and cracks appearing in the most unexpected places. Maybe it is time to view thermal differences as a potential cause of these defects.

11.9 SQUEEZED-OUT MASTIC IN WHEEL PATHS

Fat spots in wheel paths appear at the time of SMA pavement trafficking. They are located in the paths of vehicle wheels and can run up to several hundred meters (possibly including the whole SMA wearing course of a road) (Figure 11.34).

We have to go back to the chapter on designing an SMA mixture and the volume relations taking place in it to explain causes of the appearance of such "sweating offs" of mastic in the wheel paths. Recall that some air voids for mastic are intentionally left in a compacted and interlocked coarse aggregate skeleton. Additional compaction of a course under tires causes a closer arrangement of the coarse aggregate grains during trafficking. This reduces the volume of free space designed for mastic, and consequently the mastic is squeezed-out onto the surface. In some cases, this phenomenon confirms the Dutch idea about the *enlarging effect* and the need to design SMA with a bit higher void content (see Chapter 7).

Another reason for the occurrence of such a phenomenon is opening the pavement to traffic too early. Loads on an SMA that is still warm can destroy this mixture in a short time.

11.10 SMA SUSCEPTIBILITY TO POLISHING

Although susceptibility to polishing is not a problem at the construction stage but a result of an earlier decision, greater pavement slipperiness becomes apparent after some time of pavement operation. The source of this problem is at the design stage of an SMA composition. The proper selection of an aggregates for a mixture and the level of designed air voids content are crucial.

FIGURE 11.34 Mastic squeezed-out of SMA in wheel paths during trafficking of pavement. (Photo courtesy of Krzysztof Błażejowski.)

FIGURE 11.35   Polished SMA surface after 3 years of operation. (Photo courtesy of Krzysztof Błażejowski.)

We may define the expected scope of an aggregate's polishing resistance for wearing courses through the selection of the polished stone value (PSV) category. PSV checking consists of testing the microtexture loss of aggregate grains under standardized conditions according to EN 1097-8. Figure 11.35 shows the effect of polishing a pavement after 3 years of traffic. This pavement, except for polished grains, shows no other damage.

11.11   OVERDOSAGE OF AN ANTISTRIP ADDITIVE

It happens occasionally that an additive dosing system does not work correctly. When that happens, there is either too much or too little antistrip additive in the mixture. Its deficiency does not directly affect the quality of the mixture during production or placement. Effects may appear in the form of a lower durability of the course under traffic. However, an excess of the antistrip additive can manifest itself in an immediate and direct impact on the quality of the mixture. Usually the antistrip additive overdosage may be identified while still in the batching plant because of the characteristic (i.e., unpleasant) smell of the mixture.* The overdosed mixture distinguishes itself by having a very high workability both in the paver and under rollers, to such an extent that its further compaction is possible for up to a couple of days after placement (i.e., it is still deformable under rollers). Putting such a pavement into operation results in its rapid rutting. Another effect of increased workability of the mixture while rolling is the risk of fat spots or areas with a closed structure.

---

* Depends on the type of adhesive agent used (mainly concerns some fatty amines).

12 Characteristics of the SMA Course

Constituent materials, design, production, and placement of a stone matrix asphalt (SMA) mixture were discussed in previous chapters. It would not make sense to spend this much time discussing the mix if SMA courses had not been characterized with many strong points. Conversely, it should be openly admitted that it is not a perfect mixture, and it also has a couple of run-of-the-mill or slightly poorer properties among some very remarkable and even outstanding ones. In any case, its lack of perfection does not affect the final appraisal of SMA as a very useful material for pavements. After all, the rapid increase in SMA applications all over the world has not been exclusively brought about by fashion.

Next, the following operation and maintenance properties of SMA courses will be elaborated on

- Resistance to permanent deformation
- Crack resistance
- Fatigue limit
- Antinoise properties
- Antispray and antiglare properties
- Antiskid properties
- Durability
- Permeability
- Impact on fuel consumption while driving
- Economic effectiveness

This discussion will begin with a short comparison of SMA and other competitive mixtures that are used for wearing courses. Table 12.1 shows a comparison of selected properties of SMA, open graded asphalt, ultra-thin friction courses, and continuous graded asphalt mixtures (Pretorius et al., 2004). One can see from this table that SMA mixtures generally compare very favorably with the other types of mixtures but that SMA may not be the best choice for all applications.

12.1 RUTTING RESISTANCE

Resistance to permanent deformation is the best known and most recognizable feature of courses made from SMA mixtures. This resistance has its origins in the very strong skeleton of coarse particles of an aggregate mix. This issue was already thoroughly deliberated in Chapters 2, 3, and 6, therefore only some additional information about testing that feature is provided below.

TABLE 12.1

A Comparison of Essential Functional Properties of Some Popular Asphalt Mixtures for Wearing Courses

Preferred Functional Properties	Required Fundamental Mix Properties	Typical Values of Various Surfacing Types 0/13 mm Size			
		OGA	UTFC	SMA	CGA
High skid resistance	High surface texture	2–3 mm	1.5–3 mm	1–2 mm	0.2–0.5 mm
	High interconnected internal voids	18-25%	12-20%	0%	0%
	High aggregate polishing resistance	Aggregate dependent	Aggregate dependent	Aggregate dependent	Aggregate dependent
Low tire-road noise	Negative texture	Yes	Yes	Partially	No
Low spray generation	Macrotexture	2–3 mm	1.5–3 mm	1–2 mm	0.2–0.5 mm
Low construction and maintenance cost and construction delays and vehicle damage	Low layer thickness	30–40 mm	18–20 mm	30–40 mm	30–40 mm
	High durability (low maintenance cost)	Medium-high	Medium-high	High	High
	Long functional life	8–12 years	8–12 years	10–12 years	Low[a]
	Early trafficking	Yes	Yes	Yes	Yes
	Low windscreen breakages	Low	Low	None	None

Source: From Pretorius F.J., Wise J.C., and Henderson M., *Proceedings of the 8th Conference on Asphalt Pavements for Southern Africa (CAPSA'04)*, 12–16 September 2004.

Note: CGA = Continuous graded asphalt; OGA = open graded asphalt; SMA = stone matrix asphalt; UTFC = ultra-thin friction courses.

[a] Functional life is low or non-existing relative to spray reduction and wet weather friction.

12.1.1   THE IMPACT OF MIXTURE PARAMETERS

Basically, the type of aggregate mix, the type and content of binder, and the amount of air voids (Va), voids in mineral aggregate (VMA), voids filled with binder (VFB) are taken into account when analyzing the resistance to deformation of an asphalt mixture course. The development of rutting is a fairly complex process depending on, among other things, the relationships among the three aforementioned factors. Hence the resistance to deformation depends on the shear strength of an asphalt mix (Kandhal et al., 1998), which is the result of binder and aggregate (aggregate blend) interactions.

With mixtures like SMA, it would seem to be a justified statement that with such a strong aggregate structure, the role of the binder should be substantially reduced. However, practice has proved that functionally better binders like polymer modified binder (PMB) or multigrade binders are being used in more and more countries. They provide for better cohesion at high service temperatures and also improve pavement characteristics (e.g., crack resistance) at low temperatures.

Resistance to deformation has been tested for many years. Many assessment methods have been developed, from the simplest and the oldest ones like an assessment based on Marshall stability, through the so-called Marshall quotient (the ratio of stability to flow), to sophisticated up-to-date test methods including triaxial dynamic compression and shear machines.

Next, the following test methods will be discussed:

- Creep tests with constant and repeated loads
- Triaxial dynamic compression test
- Wheel-tracking test
- Asphalt Pavement Analyzer (APA)

12.1.2   Creep Tests

Encouraging results of SMA resistance to deformation are being confirmed both in laboratory tests and on trial sections of pavements in operation. Clearly, in the majority of cases an SMA mixture is actually better than asphalt concrete (AC). Conversely, not every test method confirms that SMA possesses the better characteristics. The outcome depends on the loading mode and conditions of testing. The SMA tests resulting in divergent outcomes include

- Uniaxial creep test with a constant load (also called static creep), unconfined
- Uniaxial creep test with a repeated mode of loading (also called dynamic creep or repeated load axial test [RLAT]), unconfined

The classic versions of both tests have been carried out on 100 mm diameter specimens with a 100 mm loading platen (the so-called variant 100/100).

Figure 12.1 shows the mode of loading a 100 mm diameter cylindrical specimen with a 100 mm diameter platen. The whole top face of the cylinder is being loaded in an unconfined conditions (i.e., with no side support). Bearing in mind Figure 2.3, it is obvious that the power of an SMA structure also depends on the strength of the side support (Said et al., 2000). There is no such support in an unconfined creep test and loading the sample's entire cross-sectional area. That is why, when testing SMA with this test, the results may suggest that SMA is a worse mixture than an AC mixture which has an aggregate mix characterized by a different type of particle interlocking. The fundamental assumptions of creep testing with no side support have been criticized for a long time (e.g., in Ulmgren, 1996). In spite of this, a large number of test results based on unconfined creep testing may be found in many publications.

The same tests with a different sample loading system produce different comparative results between AC and SMA. The following are two other creep tests that incorporate side support:

- Uniaxial creep test with a repeated mode of load with side support (also called indentation repeated load axial test [IRLAT])—variant 100/150
- Uniaxial creep test with a repeated mode of load with vacuum confinement (also called vacuum repeated load axial test [VRLAT]; see also triaxial test).

If a 100 mm diameter cylindrical specimen is replaced by a larger one (i.e., with one that has a 150 mm diameter [Figure 12.2]) and the loading plate has a diameter of 100 mm, then the loaded area of the specimen will be laterally supported. The

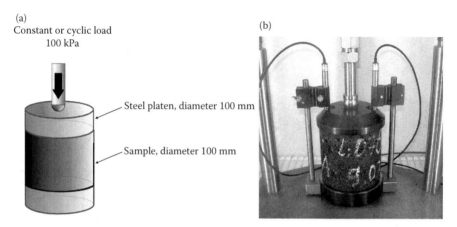

(a)
Constant or cyclic load
100 kPa

(b)

Steel platen, diameter 100 mm

Sample, diameter 100 mm

FIGURE 12.1 Creep test with no side support, variant 100/100: (a) test scheme and (b) testing in Nottingham Asphalt Tester. (Photo courtesy of Krzysztof Błażejowski.)

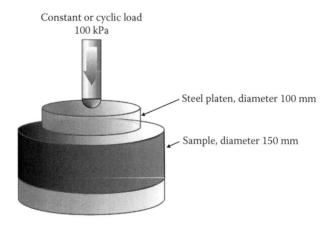

Constant or cyclic load
100 kPa

Steel platen, diameter 100 mm

Sample, diameter 150 mm

FIGURE 12.2 Mode of a modified creep test—100/150.

loading scheme (variant 100/150) and test results are, in this instance, closer to the real performance of a pavement on a road. In many countries, a modified creep test with a repeated mode of load (IRLAT or VRLAT) has been adopted for an assessment of the resistance to deformation of mixtures. Consequently, proper correlations between results of creep tests and wheel-tracking tests have been developed (Said et al., 2000; Ulmgren, 1996).

It is worth knowing that there is a substantial difference between the results of static creep tests of cylindrical specimens compacted in a laboratory for design purposes and specimens cut out of a finished course. Despite having the same composition and similar densities, specimens cut out of a course are characterized by lower values of the creep modulus than Marshall and gyratory compacted specimens prepared in a laboratory (Renken, 2000).

According to the standard EN 13108-20, static and dynamic creep tests do not apply to SMA. The cyclical compression test (100/150—IRLAT) has been described in EN 12697-25 as method A, with test parameters as follows:

- Square loading pulse
- Frequency of 0.5 Hz (load 1 sec, rest period 1 sec)
- Load of 100 ± 2 kPa
- Typical test temperature of 40°C
- Total test duration of 3600 cycles (2 hours)

12.1.3 TRIAXIAL DYNAMIC COMPRESSION TEST

A concept similar to side supported creep tests applies to the triaxial dynamic compression test. In this method, a specimen is subjected to compression with lateral support. This method has been widely regarded as one of the most accurate, reflecting the state of stress in a loaded pavement. Moreover, it enables the measurement of parameters of an asphalt mixture used for the analysis of pavement viscoelasticity (Huurman, 2000). Undoubtedly, it is a recommended method for testing rut resistance. It is described in EN 12697-25 as method B.

12.1.4 WHEEL-TRACKING TEST

The best known direct tests for resistance to deformation are wheel-tracking tests, conducted with special devices. There are many types of them, such as European devices (according to EN 12697-22), the Hamburg Wheel-Tracking Device, and the APA.

The European standard EN 12697-22 classifies the rutting equipment into small devices and large devices.

According to Table B.5 in EN 13108-20 (the standard that regulates the methods for asphalt mixture type testing), the small device method (testing in air) has application in testing SMA. It is meant for pavement courses subjected to loads less than 13 tons per truck axle. Pavements subjected to loads heavier than 13 tons per axle are tested in a large device. More information on that subject can be found in Chapter 14.

12.1.5 ASPHALT PAVEMENT ANALYZER

The APA is the second generation of Georgia loaded-wheel tester (GLWT) used in the United States for testing resistance to deformation of asphalt mixtures. The test temperature depends on the climatic data for the region where the mixture will be placed, and it is usually close to the highest expected temperature of the pavement. The test conditions include the wheel load and contact pressure, which are individually determined (usually 445 N and 690 kPa, respectively), and the number of loading cycles, 8000.

A full report about the APA can be found in the U.S. publication *National Cooperative Highway Research Program* Report No. 508 (Kandhal and Cooley, 2003).

12.1.6 TEST RESULTS

Rutting resistance is one of the most widely tested properties of asphalt mixes, including SMA, and various methods of testing are conducted all over the world.

The impact of side support during creep tests of SMA and AC have been compared in Swedish research (Said et al., 2000). Results have explicitly shown the great significance of that feature to SMA, while the results from testing AC with and without side support have not changed considerably (Figure 12.3). Other research (Ulmgren, 1996) on the comparison of the dynamic creep test (RLAT 100/100 mm) and the modified one (IRLAT 100/150 mm) with the results of a wheel-tracking test have demonstrated a very good relationship between IRLAT and the wheel-tracking test, with R = 0.91, while the relationship between RLAT and the wheel-tracking test was much worse, with R = 0.36.

U.S. research (Cooley and Brown, 2003) on SMA mixtures used for thin wearing courses has proved the advantages of conforming with requirements for resistance

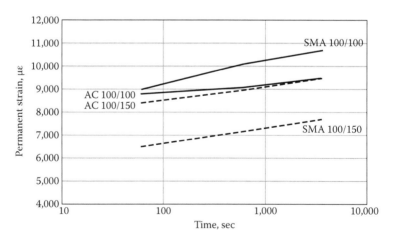

FIGURE 12.3   Comparative results of SMA and AC with and without side support during the creep tests RLAT 100/100 and IRLAT 100/150. (From Said, S., Jacobson, T., Hornwall, F., and Wahlström, J., *Proceedings of the 2nd Eurasphalt & Eurobitume Congress*, Barcelona, 2000. With permission.)

to deformation (based upon APA testing), despite the maximum gradation of the SMA not exceeding 9.75 mm. British results (Obert, 2000) concerning the comparison of SMA with hot rolled asphalt (HRA) have confirmed a higher resistance to deformation more for the former than the latter. Other research undertaken in Finland on test sections (Kelkka and Valtonen, 2000) compared SMA with AC, graded 0/18 mm with various binders. This time SMA turned out to be the winner again. Polish research (Sybilski and Horodecka, 1998) has acknowledged that creep tests are not appropriate for determining SMA resistance to deformation and also that better SMA rutting resistance has been proved in comparison with AC on road sections and in wheel-tracking testing (large device—French LCPC type).

Briefly, to sum up this short review of test results with regard to rutting resistance, it may be concluded that they reflect SMA's superior properties over those of AC.

12.2   CRACK RESISTANCE

Low temperature cracking and reflective cracking will be discussed next. More information on the theory and origin of cracking can be found in the literature (Arand, 1996; Jacobs et al., 1996; Rigo, 1993).

12.2.1   Low Temperature Cracking

Low temperature cracking induced by a drop in temperature has been well-documented (e.g., Fabb, 1973; Isacsson et al., 1997; Marasteanu et al., 2004; Tuckett et al., 1969). Cracking of an asphalt course appears when the thermal stress, which increases with a drop in the temperature, exceeds the mixture's tensile strength. It originates at the surface of the wearing course and advances downward.

An overview of factors influencing the development of low temperature cracking has been presented in the literature (Isacsson et al., 1997). Although binder properties have commonly been regarded as responsible for a pavement's susceptibility to this type of cracking, there are actually many other, though less common, contributing factors (e.g., the content of voids and mastic).

12.2.2   Reflective Cracking

Reflective cracking is a well-known weak point of semirigid pavements. It advances upward from a rigid base through the asphalt layers. Various techniques for countering reflective cracking (e.g., Stress Absorbing Membrane Interlayer (SAMI) and Stress Absorbing Membranes (SAM) membranes, geogrids) have been used. But their effect usually comes down to more or less effectively slowing the growth of cracking, not preventing it from occurring. Asphalt courses with increased crack resistance are characterized by considerable shear and tensile strength. These properties may be achieved through the appropriate selection of the gradation of an aggregate mix, the type of binder and possibly the mastic strengthening additives.

12.2.3 TEST METHODS FOR CRACK RESISTANCE

Different methods for testing crack resistance have been used, and so far there has not been one commonly regarded as the dominant standard. There are some popular methods such as the thermal stress restrained specimen testing (TSRST; discussed later), local procedures used by specific research centers (e.g., Judycki, 1990), and methods under standardization, such as the semicircular bending test (Krans et al., 1996; Molenaar and Molenaar, 2000). Test methods for asphalt binders (e.g., the BBR method) have also opened up some new possibilities. In addition, the oldest engineering method—namely, observation of test road sections—is still in use.

12.2.3.1 Thermal Stress Restrained Specimen Test Method

The TSRST method consists of attaching the ends of a specimen (250 mm long and 60 mm in diameter or 50×50 mm cross section) to the frame of a device situated in a cooling chamber. The frame is rigid to keep the length of the specimen unchanged. As the temperature falls (at a rate of $-10°C/hr$), the specimen contracts and the tensile stress in the specimen rises because it is being held by the frame. The temperature at which the specimen cracks is the test result. This method has been standardized in AASHTO TP 10 and prEN 12697-46.

12.2.3.2 Semicircular Bending Test

The semicircular bend test method is described in prEN 12697-44. A half cylinder sample of compacted asphalt mixture is loaded using a three-point bending scheme. As a result, tensile stress is created at the bottom of sample. This test can be used either for testing fracture toughness (when the half cylinder has a crack sawed at the bottom center with a notch width of 0.35 ± 0.10 mm and a depth of 10 ± 0.2 mm) or for tensile strength (unnotched cylinder).

12.2.4 TEST RESULTS

A significant impact of mixture type on cracking temperature for newly made mixtures or after a short time of aging has not been observed with TSRST testing mixtures of AC, SMA, and porous asphalt (Isacsson et al., 1997). Not until a longer aging period (i.e., more than 25 days) has elapsed has a difference in favor of AC and SMA been observed. Porous asphalt is more susceptible to aging, and the recorded difference of crack temperatures has reached 6°C and 25°C after 25 and 100 days of aging, respectively. Test results (Judycki and Pszczoła, 2002) concerning a comparison of low temperature properties of various mixtures have not revealed essential differences between AC and SMA. In a U.S. site investigation (Schmiedlin and Bischoff, 2002), the comparison of capabilities to slow down the advancement of reflective cracking in various mixtures concluded that SMA is superior to AC. The impact of aggregate size and its resistance to crushing has also been noted. Those conclusions are based on a 5-year observation of test sections. Similar results have been achieved on a test section in Australia (Pashula, 2005) where, among other things, the abilities of various mixtures to slow down the advancement of reflective cracking were compared. It was stated after a 10-year observation that SMA's distinctive feature is that it possesses the greatest potential of slowing the occurrence of reflective cracking.

12.3 FATIGUE LIFE

Fatigue is an effect consistent with the formation of cracks in material caused by a series of repetitive tensile stress cycles that do not exceed the tensile strength of the material. (For more information on fatigue, refer to the many publications with descriptions of this phenomena [e.g., SHRP Reports A-312 or A-404].)

12.3.1 TEST METHODS

There are many methods for testing fatigue; for example, the European standard on testing fatigue EN 12697-24 has quoted the following ones:

- Two-point bending test on trapezoidal-shaped specimens (2PB-TZ)
- Two-point bending test on prismatic-shaped specimens (2PB-PR)
- Three-point bending test on prismatic-shaped specimens (3PB-PR)
- Four-point bending test on prismatic-shaped specimens (4PB-PR)
- Indirect tensile test on cylindrical-shaped specimens (IT-CY)

Note: the standard EN 12697-24 has clearly stipulated that results obtained with various methods are not comparable; also, the standard EN 13108-20 has limited fatigue tests exclusive to AC mixtures (as a part of initial type testing).

Extended comparisons of fatigue test methods can be found in the literature (di Benedetto et al., 1997; Said and Wahlström, 2000).

Fatigue tests are carried out under one of the following modes of loading:

- Stress controlled
  - Stress is induced in a specimen and is held throughout the test; strain steadily increases with the loading cycles until failure of the specimen occurs, which signals the end of testing.
  - The fatigue limit is proportional to the mixture's stiffness.
- Strain controlled
  - Strain is induced in a specimen and is held throughout the test; stress steadily decreases with the application of loading cycles until the specimen's stiffness reaches 50% of its initial level.
  - The fatigue limit is inversely proportional to the mixture's stiffness.

12.3.2 TEST RESULTS

A fairly comprehensive range of methods and test parameters can be found in publications on SMA fatigue properties. For example, in Australian research (Stephenson and Bullen, 2002) the strain-controlled mode was used to conduct a four-point bending beam test at 20°C, with continuous haversine loading at a frequency of 10 Hz and a range of strain levels from 100 to 1000 $\mu\epsilon$. On the grounds of the test results, it can be stated that the fatigue limit of an SMA mixture is higher than a comparative specimen of AC.

12.4 WORKABILITY

The concept of workability has been used for determining a series of mixture properties significant at the time of placement of a pavement. Workability is the property that determines the ability of a mixture to be placed mechanically or spread manually and finally compacted (Asphalt Institute Handbook MS-4, 1989; Gudimettla et al., 2003). Naturally, compactability is a feature of less extensive significance, so it is a reflection of workability.

Obviously, workability is affected by the content and type of binder and the mixture temperature. It has been stated in U.S. research (Gudimettla et al., 2003) that workability is also influenced by the properties of the aggregate mix and the maximum particle size.

12.5 COMPACTABILITY

12.5.1 Definitions and Test Methods

The notion of compactability has been repeatedly used in this text, especially in Chapter 10. Compactability (i.e., susceptibility to compaction) can be defined as the ability of an asphalt mixture to change density under the influence of compactive effort; or to put it another way, compactability is the material feature determined by the amount of energy necessary for the compaction of a given mass into the smallest volume (Schabow, 2005). Resistance to compaction is the reverse of compactability. Generally, it means that compactable mixtures (with a low resistance to compaction) do not need a lot of compactive effort.

Resistance to compaction is connected with features of a mixture, such as the gradation of the aggregate mix and the aggregate properties, which include the following:

- Content of crushed stone
- Particle microtexture
- Particle shape
- Hardness (resistance to crushing and wearing)

Additionally, resistance to compaction is also affected by the content and type of binder and its viscosity at the compaction temperature.

Compactability is tested in Europe in accordance with the European standard EN 12697-10, which provides for the following two methods for testing and assessing the compactability asphalt mixtures:

- Method I—a series of specimens are compacted with different efforts, and their bulk densities are determined.
- Method II—based on a height measurement, a change (i.e., an increase) in the bulk density is determined after completing each stage of compaction.

The following equipment can be used for testing compaction resistance and compactability according to EN 12697-10:

- Impact compaction (Marshall hammer) according to EN 12697-30
    - Method I—results as compaction resistance C (units [42 Nm])
    - Method II—results as compaction resistance T (units [21 Nm])
- Gyratory compaction according to EN 12697-31
    - Method I—not used
    - Method II—results as compactability K (dimensionless)
- Vibratory compaction according to EN 12697-32 (not applicable to SMA)
    - Method I—results as compactability k (dimensionless)
    - Method II—not used

12.5.2 RESULTS OF SMA COMPACTABILITY TESTS

SMA is a hard-to-compact mixture, and as a gap-graded mixture, it is characterized by a higher compaction resistance than materials with continuous gradation. The conclusions of tests carried out in Germany (Renken, 2004) on an SMA mixture revealed the following:

- An increase of the filler content brings about a drop in compaction resistance.
- An increase of the coarse aggregate content (particles larger than 2 mm) causes a sharp rise in compaction resistance.
- An increase of the binder content slightly reduces the compaction resistance.

Additionally, practice has proved that an increase in the manufactured sand content results in an increase in compaction resistance and that the opposite is true for the quantity of binder (those are commonly known relations). SMA mixtures with a low content of voids (1.5–2.5% v/v) are compacted more easily than those with a content greater than 4% v/v (Schroeder and Kluge, 1992).

When producing an SMA mixture, substantial deviations in batching the filler, binder, and coarse aggregate fraction also promote changes in compaction resistance, in addition to alterations in other properties.

12.5.3 SMA COMPACTION ENHANCING AGENTS

Agents enhancing compactability by means of changing the temperature susceptibility of a binder have been used in many countries. They enable the placement of a mixture at a lower temperature and make its compaction easier through the reduction of binder viscosity. Reduced binder viscosity enhances the compactability of a mixture, resulting in a decrease in the content of voids and an increase in the bulk density. In fact, it is this effect that can be seen in Figure 12.4, which shows example test results of a binder containing a Fischer–Tropsch (FT) wax. The application of an agent of this kind causes an increase in the bulk density of a mixture by about 15% in comparison with mixtures without this agent compacted at the same temperature. The presented mixtures contained the polymer modified binder PmB 45A (German designation).

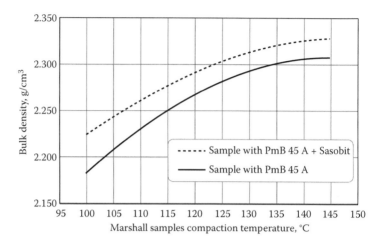

FIGURE 12.4   Impact of FT wax on SMA compactability. (From Damm, K.W., *Journal of Applied Asphalt Binder Technology*, 2, 2002. With permission.)

12.6   ANTINOISE PROPERTIES

Noise had been defined as "audible sounds of any acoustic kind undesired in particular circumstances, which irrespective of their frequency and level, are harmful, bothersome, and possibly induce a disorder in the listener's hearing organ and other parts of their organism" (Kucharski, 1979). Many research centers around the world have been dealing with the problem of noise, and numerous publications have dealt with the subject. A comprehensive review of publications addressing the subject of noise can be found in Sandberg and Ejsmont (1999).

A source of noise emits an acoustic wave, which is subject to reflection and partial absorption by the pavement. "Silent" pavements are those with reasonably high sound absorption capabilities. Absorption depends on the characteristics of the pavement surface and the shape of available air voids.

12.6.1   Test Results

There are a number of impressive publications in the technical literature that describe the results of testing antinoise properties of various pavements. (Refer to the Bibliography at the end of this book.)

Olszacki (2005) tested the sound-absorbing power of different asphalt surfacing types with diversified void contents. Figure 12.5 shows the relationship between the noise absorption coefficient and sound frequency. It is evident that SMA is characterized by better properties than mixtures of AC, but not as good as porous asphalt with a much higher content of voids, from 10–22% (v/v).

Other research has also stated that pavement noise increases along with an increase in the maximum particle size of the wearing course mixture. Therefore, when antinoise properties are at issue, mixtures SMA 0/5 and 0/8 instead of 0/11 and 0/16 mm are preferred. Generally, the macrotexture of SMA makes it quieter

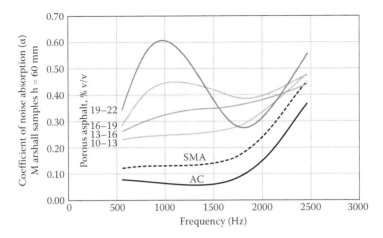

FIGURE 12.5   Impact of asphalt mixture type on noise absorption. (From Olszacki, J., The determination of the water permeability and noise absorption of asphalt concrete used in porous courses [Określenie wodoprzepuszczalności i dźwiękochłonności betonoasfaltów stosowanych w nawierzchniach drenujących], Ph.D. thesis, Kielce University of Technology, Poland, 2005. With permission.)

than AC by about 1–2 dB(A). When coarse SMA 0/16 is used, an increase in the noise level of about 1 dB(A) is reported (Sandberg, 2001). SMA's distinctive feature—namely, the grit of 2/4 or 2/5 mm aggregates—makes the pavement noisier, which is why gritting with finer aggregate is recommended. After some time, when the grit been removed by traffic, the SMA noise level will naturally reduce. So the conclusion may be drawn that the SMA antinoise properties change over the service life of the pavement. A site investigation (Schmiedlin and Bischoff, 2002) has proved that a classic SMA pavement is only slightly quieter than classical AC.

Finnish tests of SMA antinoise properties (Valtonen et al., 2002) have pointed out another problem when studded tires are permitted. Testing using the close-proximity (CPX) method was conducted at a speed of 50 km/hr. In spite of the SMA 0/5 mm having the best properties just after laydown (in comparison with SMA 0/8, 0/11, and 0/16), a deterioration in properties due to wear caused by studded tires was found after the first winter of operation (only 1 year of service life). Under these circumstances, the SMA mixture's resistance to wear by studded tires should have been taken into account. The coarsest SMA gradation is more resistant to studded tire wear, in contradiction to its antinoise properties. Consequently, it has been concluded that SMA 0/8 and SMA 0/11 might be more appropriate than finer SMA 0/5 (in this research, SMA 0/5 has worn 10 times as much as SMA 0/16). In testing done in the United States, SMA 0/12.5 and SMA 0/9.5 mm were compared using the CPX method (Bennert et al., 2004). The results proved the increase in noise at the contact between tires and pavement along with the increase of SMA maximum aggregate size.

Table 12.2 shows the collective comparison of SMA properties with a reference mixture of AC. The new concept of a silent SMA characterized by better properties than classical SMA is presented in Chapter 13.

TABLE 12.2
Noise Levels of SMA Compared with Other Asphalt
Mixture Types

Country	SMA Type	Reported Reduction dB(A)[a]	Reference
Germany[b] (v = 50 km/h)	0/5, 0/8	+ 2.0 to –2.0[c]	BA 0/11
Italy (v = 110 km/h)	0/15	+ 5.0 to + 7.0	BA 0/15
The Netherlands (v = 60–100 km/h)	0/6	+ 1.4 to + 1.6	BA 0/16
	0/8	+ 0.2 to + 0.6	
	0/11	0.0 to –2.0[**]	
		+ 0.8 to –0.5	
		+ 1.0 to –3.0[**]	
United Kingdom (v = 70–90 km/h)	0/6	+ 5.3 to + 5.2	Hot Rolled
	0/10	+ 3.5 to + 3.2	Asphalt
	0/14	+ 2.7	

Source: From EAPA, Heavy duty surfaces. The arguments for SMA. European
   Asphalt Pavements Association (EAPA), 1998, With permission.
[a] Negative values indicate an increase in noise level.
[b] Calculated value.
[c] When the surface is treated with uncoated chippings smaller than 2 mm.

12.7   ANTISKID PROPERTIES

Antiskid features of wearing courses of asphalt surfacing are described in terms of the friction coefficient, which depends on the micro-texture of the aggregate and the macro-texture of the placed mixture (Gardziejczyk and Wasilewska, 2003). As this property is very important, the range of available publications on the topic at issue is quite broad (Gardziejczyk, 2002; German DAV Report, 2001; Huschek, 2004; Jordens et al., 1999).

Different pavement qualities, in relation to the speed of vehicles, have a decisive impact on antiskid properties (Hunter, 1994). At low speeds, polishing resistance has a decisive impact, hence the durable microtexture of the aggregate surfaces is significant; in this instance the polished stone value (PSV) can be used as a selective index for evaluating an aggregate's resistance to polishing. At high speeds, the macrotexture depth of the wearing course has a decisive impact; the presence of small channels around aggregate particles on the surface enables water discharge, preventing the formation of a hydroplaning effect (skidding layer).

The SMA macrotexture depth depends on the maximum aggregate size in the mixture and the design of the mix (e.g., the level of filling voids among coarse aggregates on the surface of a course with mastic); see Figure 12.6.

(a)   (b)

FIGURE 12.6 SMA macrotexture before opening to traffic: (a) nongritted and (b) gritted. (Photo courtesy of Krzysztof Błażejowski.)

12.7.1 TEST RESULTS

A comparison of macrotexture depths of courses executed with various asphalt mixtures has been presented in an Australian paper (Oliver, 2001):

*   Surface dressing (greater than 10 mm)—macrotexture less than 1.5 mm
*   AC (greater than 10 mm)—macrotexture 0.4 to 0.8 mm
*   Porous asphalt—macrotexture less than 1.2 mm
*   SMA—macrotexture less than 0.7 mm
*   Slurry seal—macrotexture 0.4–0.8 mm
*   Cement concrete (brushed)—macrotexture 0.2–0.7 mm

Changes in some SMA characteristics occurring over time have been indicated in various investigations. In the United Kingdom, Richardson (1997) described changes in the SMA macrotexture during the months following the construction of a course. The SMA macrotexture, just after placement, was initially at a level of 1.5–1.6 mm. Then it gradually reduced to 1.1 mm after 21 months. It should be added that the reported results applied to an SMA with a gradation that was used in the United Kingdom to achieve a required high macrotexture after placement (see Chapter 10). The applied gradation was characterized by a gradation curve below the typical gradation limits according to German guidelines (see comparison of curves in Chapter 14, Figures 14.2 through 14.5).

It should be noted that gritting used to enhance SMA antiskidding properties has additional consequences, especially in the first period after execution, namely a slight decrease in the surface macrotexture. Grit particles (and the remains of them crushed by rollers) gather in spaces among SMA coarse aggregates and only some time after the opening of a road to traffic are they pulled up (sucked out) by vehicle tires and scattered across the shoulder or median (Richardson, 1997). So just after gritting, macrotexture may be low (see Figure 12.6b), it but will eventually improve.

Argentinean investigations (Bolzan, 2002) conducted just after a highway modernization project, including the placement of a new SMA 0/19 course, showed it provided high macrotexture 2.2 mm deep (measured by the sand patch method). At another section of SMA 0/12 mm, the macrotexture depth amounted to only 1.4–1.7 mm.

In Germany Behle et al. (2005) conducted research into the relationship between PSV and final skid-resistance of an SMA layer. The authors presented the results of measurements with a sideways coefficient routine investigation machine (SCRIM) over a period of 4.5 years and a concept for calculating the PSV of mixed aggregates with different individual PSVs by weighted average.

By and large, based on straightforward experiments and a series of results, the following general technological recommendations to improve the antiskidding properties of SMA wearing courses may be put forward:

- Apply mixtures with a lower maximum aggregate size (more contact points between the SMA and tire).
- Consider microtexture—prefer aggregates with a high PSV index (low polishing susceptibility) and aggregate mixtures of various rocks with various wear rates; if possible, also use artificial aggregates (slags).
- Consider macrotexture—avoid factors that increase the risk of squeezing mastic out on a surface such as mixtures with an insufficient content of voids and or those susceptible to compaction under traffic; for the same reason do not use pneumatic rollers and use vibratory rolling with caution.
- Apply grit to make the surface rough by spreading the grit evenly, followed by rolling while the surface is hot enough.
- Open the lane to traffic only after the SMA has finally cooled off.

12.8 DURABILITY: WATER AND FROST RESISTANCE

Pavement durability is a broad term. Water and frost resistance of asphalt mixtures have a disadvantageous effect on the mechanical performance of a course. Undoubtedly, the composition of a mixture—the type of aggregate, gradation of the mix, the type and quantity of binder, the presence of additives and the content of air voids—has an impact on this resistance. More information on this issue may be found in different publications (e.g., Kanitpong and Bahia [2003]; Santucci [2002])

The most common assessment methods for water and frost resistance of asphalt mixtures can be divided into the following two groups:

- Standardized methods
  - Method AASHTO T 283
  - Method EN 12697-12
- Non-standardized methods (devised in research centers for particular or local use)

A description of other methods for testing durability of asphalt mixtures and more details of this subject can be found in various papers (Chen et al., 2004; Hicks et al., 2003; Judycki and Jaskuła, 1999; Martin et al., 2003; Ulmgren, 2004).

12.8.1 AASHTO T 283 METHOD

The AASHTO T 283 method involves conducting tests on a comparable set of specimens in an original (unconditioned) state and after conditioning and then comparing the results. In some literature this test is also called the modified Lottman test.

Appropriately prepared specimens are divided into two sets. One set is designed for testing without conditioning, while the other is subjected to conditioning in water and freezing. Both the original and conditioned specimens are tested with an indirect tension apparatus. The ratio of the conditioned tensile strength to the strength of the original specimens is called tensile strength ratio (TSR). TSR is usually required to be greater than 70% or 80% (most often 80%).

12.8.2 EN 12697-12 METHOD

This EN 12697-12 method test is conducted on cylindrical specimens prepared in a laboratory (in a gyratory compactor, using a Marshall hammer) or cored from a slab cut out of a pavement. Specimens of 100 ± 3 mm, 150 ± 3 mm, or 160 ± 3 mm in diameter may be tested. When testing specimens 100 mm in diameter (compacted using a Marshall hammer, for instance), only mixtures with a gradation not larger than 0/22 mm can be tested. The set of specimens (minimum of six) of an asphalt mixture is divided into two groups. Both groups should be prepared at approximately the same time (within 1 week or less of each other). Half of the specimens are stored without conditioning, while the other half are subjected to conditioning in water. Specimens are compressed in an indirect tensile test using EN 12697-23. The compression test may be conducted at a selected temperature within a range from 5–25°C.

The standard introduced an index called the indirect tensile strength ratio (ITSR) as a measure of the water resistance of an asphalt mixture, expressed in percent. Apart from the ITSR index, the type of failure, the degree of coating of an aggregate with a binder in a brittle fracture, and the type of aggregate breakage should be given in a test report.

12.8.3 NONSTANDARDIZED METHODS

These methods are normally based on similar assumptions—namely, in saturating a mixture with water (with or without negative pressure, that is a vacuum) and holding it there at a fixed temperature for a given time. Afterward, a strength test is conducted, most often using one of the following methods:

- Marshall stability
- Resilient modulus at different temperatures
- Indirect tensile strength

The comparison of results for specimens conditioned in water with those untreated in water determines the water resistance of an asphalt mixture.

In an extended variant, specimens saturated with water are subjected to many cycles of freezing and thawing to find the mixture's susceptibility to water and frost. Another variant that involves freezing specimens previously saturated with an aqueous solution of NaCl (e.g., 2%) has also been used. This is a much more effective test due to the aggressive action of the aqueous solution of salt on binder adhesion to the aggregate. This kind of test is often conducted in countries with colder climates.

12.8.4 Test Results

Durability tests of asphalt mixtures, especially those intended for wearing courses, have become commonly used around the world (Höbeda, 2000; Sybilski and Mechowski, 1996). The above-average durability of SMA has been emphasized in numerous publications so they will not be described here. This durability is the result of a high binder content and a thicker binder coating on an aggregate that has the same content of air voids as AC.

12.9 ANTISPRAY AND LIGHT REFLECTION PROPERTIES

The safety provided by wearing course mixtures is of primary importance. The anti-spray properties of some mixes enhance safety by preventing the buildup of water mist behind vehicles. After rainfall (or melting snow), drops of water are raised up from a pavement by vehicle tires, especially the tires of large, fast-moving vehicles. Consequently, a certain amount of water in the form of mist remains suspended in the air surrounding fast-moving vehicles (Figure 12.7). Then it settles on windscreens, reducing visibility.

The generation of water mist and splash may be decreased by enabling quick water discharge after rainfall (through transverse cross-falls and having no ruts to hold water), a suitable depth of macrotexture, or a high content of air voids in a course. The last condition refers to open-graded friction course or porous asphalt. Water discharge through the whole thickness of a course is out of the question in the case of SMA. It is possible only through spaces among coarse aggregates sticking

FIGURE 12.7   Water mist picked up by vehicles driving along a wet SMA pavement is reduced. (Photo courtesy of Krzysztof Błażejowski.)

up from the surface of the course; hence, the proper SMA macrotexture is all important (Figure 12.8). Investigations conducted in the United States have proved that SMA courses reduce water splash in comparison with AC pavements; nevertheless, water remains longer on the SMA course where it can be held in the surface voids (Schmiedlin and Bischoff, 2002). The issues of SMA macrotexture were discussed in the Section 2.7 devoted to antiskidding properties.

Furthermore, macrotexture that reduces the water film on the surface of a wearing course is also of significance to the improvement in visibility of road marking after or during rainfall. Moreover, at night the light reflection ("glistening") of vehicles travelling in the opposite direction is reduced. Figure 12.9 shows an image

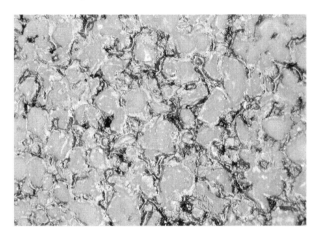

FIGURE 12.8   Water discharge through small channels around SMA coarse aggregates. (Photo courtesy of Krzysztof Błażejowski.)

FIGURE 12.9   Differences in pavement abilities to discharge water—the SMA course *(near)* and the old dense asphalt concrete *(far)*. The light reflection off the wet SMA course is substantially less than that off the asphalt concrete course. (Photo courtesy of Krzysztof Błażejowski.)

of wearing course surfaces made of SMA and AC after rainfall and their light reflections (during the daytime). The difference in the way water is discharged as a result of the surface macrotexture is clearly noticeable.

12.10 PERMEABILITY

For many people the surface appearance of an SMA course gives the impression of being excessively water permeable. Indeed, SMA differs from AC due to a substantially deeper macrotexture. SMA courses with an increased content of voids may occasionally be seen in practice, most frequently when the mixture was improperly compacted on a work site (such cases were elaborated on in Chapter 11). However, there is more to permeability than work site error.

Permeability is related to the content of air voids and the size, distribution, and existence of interconnections between internal pores. The permeability of a mixture depends on the maximum aggregate size, the thickness of the course, and the level of compaction (WsDOT, 2005). Research has shown that the larger the maximum aggregate size in a mixture, the higher its permeability (Cooley et al., 2002), which means an increase in maximum aggregate size results in the growth of the pore sizes and an increased probability of their connection (Cooley and Brown, 2003). The influence of VMA is also seen; permeability decreases as VMA increases for constant air voids (Brown et al., 2004).

To reduce the permeability of courses, one should appropriately match the mixture gradation to the course thickness so that the ratio of thickness to the maximum aggregate size in a mixture is not less than 3.0 (a ratio closer to 4.0 is preferred for SMA with a strong gap gradation). This should make the compaction of the course easier, reduce the content of voids, and limit connections among pores (Cooley et al., 2002). The following list summarizes the important points regarding permeability:

- The larger the maximum aggregate size in a mixture, the higher its permeability.
- The higher the content of air voids in a course, the higher its permeability; however, mixtures with the same content of voids can have different levels of permeability.
- The gradation of the SMA mixture (more or less gap graded) influences the permeability.
- The thicker the course, the lower its permeability.

12.10.1 TEST RESULTS

Figure 12.10 shows the results from tests of water permeability of an SMA mixtures with gradations 0/4.75, 0/9.5, and 0/12.5 mm (Cooley and Brown, 2003). All instances concern U.S. SMA, previously described in Chapters 6 and 7.

Figure 12.10 clearly shows the relationship between the maximum aggregate size of the SMA and the probability of its being permeable. The larger the maximum aggregate size of the SMA, the higher the probability of its being permeable. Thus the size of the maximum particle is a decisive factor for permeability of an SMA

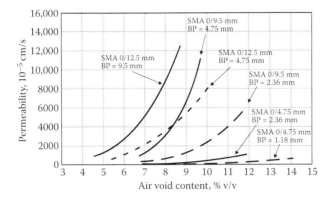

FIGURE 12.10 Water permeability of SMA mixtures with gradation 0/4.75, 0/9.5, and 0/12.5 mm, depending on the size of the selected breakpoint sieve—the degree of gap gradation. (From Cooley, L.A. Jr. and Brown E.R., Potential of Using Stone Matrix Asphalt [SMA] For Thin Overlays. National Center for Asphalt Technology, Auburn University, NCAT Report 03-01, April 2003. With permission.)

mixture and so is the gradation within the coarse aggregate fraction (driven by the breakpoint sieve using the U.S. definition). An increase in this factor is followed by the growth of the size of internal pores, and consequently, the probability of their connection (Cooley and Brown, 2003). It has been stated in research (Cooley et al., 2002) that SMA mixtures are characterized by a higher potential for permeability than AC mixtures with the same content of air voids. Investigations of permeable pavements in Florida in the United States led to a definition of permeable mixtures as those mixtures with field permeabilities greater than 100×10^{-5} cm/sec[*] (Choubane et al., 1998). SMA mixtures may be prone to this excessive permeability, particularly those with a nominal maximum aggregate size greater than 10 mm.

12.11 FUEL CONSUMPTION

The European Asphalt Pavements Association and Eurobitume report of 2004 entitled *Environmental Impacts and Fuel Efficiency of Road Pavements* (Beuving et al., 2004) is one of the main sources forming the basis for this section of the present chapter.

Rolling resistance is one of the numerous factors of intense interest when considering the problem of fuel consumption while driving, especially in the context of SMA. Rolling resistance may be defined as the force necessary to move a vehicle along a pavement.

At a constant speed of 80 km/h, approximately 12% of the energy loss (fuel consumption) of a heavy truck is consumed in overcoming the rolling resistance; the energy spent on overcoming this resistance equals about 30% of the potential available mechanical power at the engine crankshaft (Sandberg, 2001).

---

[*] Limit suggested when evaluating the in-place superpave mix pavement permeability.

Briefly, the rolling resistance is affected by microtexture, macrotexture, megatexture, and the unevenness of a wearing course.

According to Dutch research (Roovers et al, 2005), rolling resistance can be ranked by type of wearing course (results in cR[%]) as follows:

Cement concrete with burlap (smooth)—0.86
SMA 0/8—0.86
Double layer porous asphalt 2/6—0.97
Double layer porous asphalt 4/8—1.02
Cement concrete transversely brushed (rough)—1.04
Single layer porous asphalt 6/16—1.05
Dense AC 0/16—1.09
Mastic asphalt 0/11—1.18

Finally one should remember that rolling resistance measurements are strongly influenced by weather (e.g. sidewind velocity). Weather conditions could affect the test results.

Swedish investigations (Sandberg, 2001) have additionally pointed out that the pavement unevenness increases fuel consumption by as much as 12%, which seems to be a more significant factor than the type of asphalt surfacing.

12.12 ECONOMICAL EFFECTIVENESS

In most countries, SMA mixtures are more expensive to construct than comparable mixtures of AC. Higher initial prices result from the application of the following:

- Larger amounts of binder (or a PMB)
- Greater amounts of added filler
- Large quantities of high quality coarse aggregates
- Stabilizers (most often fibers)
- Higher production temperatures
- Lower outputs of asphalt plants

The approximate price difference amounts to + 20% to + 30%, depending on the country and specificity of the placement. However, such a difference in price is accepted by road administrations owing to the better durability of SMA pavements. It is widely assumed that their average lifetime amounts to at least 20 years. In many countries it is difficult to verify this service life due to the small number of SMA sections that have been in place longer than 15 years.

It can be assumed that the higher initial costs of SMA mixtures have been compensated for by their longer durability and lower maintenance costs. Taking into account the lower costs of operation due to the absence of repair needs and hence fewer traffic disruptions for road users, the economical efficiency of SMA is higher than that of the classic AC.

13 Special Applications of SMA

This chapter describes some special applications of stone matrix asphalt (SMA). Several brief case studies illustrate some of the less common, though advantageous, ways SMA mixtures can be used and also suggest some areas where they should be used with caution.

13.1 AIRFIELD PAVEMENTS

According to a report by the European Asphalt Pavement Association (EAPA), asphalt surfacing covers the majority of runways (EAPA, 2003). The high performance of SMA pavements has been attracting the attention of airport management, creating the chance to apply SMA technology in wearing courses of airfield pavements. Numerous trial sections have sprung up on various airfields (e.g., Sydney, Australia, and Johannesburg, South Africa). Some important, larger applications of SMA may be seen in Frankfurt on the Main, Germany, and Gardermoen near Oslo, Norway. The airfield in Frankfurt will be described in detail later on, not only for the SMA technology itself but for its application.

13.1.1 REQUIREMENTS FOR AIRFIELD PAVEMENTS

Requirements concerning wearing courses of airfield pavements are defined as follows (namely, in the regulations of the U.S. Federal Aviation Agency (FAA) Advisory Circular AC No. 150/5370-10B):

- Be impermeable to water and suitably protect intermediate course.
- Provide surface free of foreign object damage (FOD) (i.e., loose particles).
- Resist loads from aircrafts.
- Be smooth, with uniform surface.
- Maintain required antiskid properties.
- In specific areas should be resistant to spillage of fuel, hydraulic and deicing fluids, or other solvents.

Guidelines on specifications for pavements on civil installations are contained in *International Standards and Recommended Practices, AERODROMES. Annex 14 to the Convention on International Civil Aviation, Volume I—Aerodrome Design and Operations* by the International Civil Aviation Organization (ICAO). The ICAO regulations regarding civil aerodromes mainly define the features of pavements affecting air traffic safety—minimum level of friction, sufficient smoothness, lack of presence of any loose particles (i.e., FOD) or grains bigger than 3 mm resting on the pavement.

The rules of the FAA and ICAO govern the methods and frequencies of testing the friction and macrotexture. Methods of permanent measurement of the friction factor described in both documents will not be referred to here. Measurements of macrotexture are usually carried out either with calibrated sand (sand patch test) or with the application of special greases (NASA grease smear method). The minimum surface texture depth recommended by the ICAO is 1.0 mm. It is worth noting here that the essential task of macrotexture is to enable water discharge from the pavement so that a layer of water does not build up between a wet runway and an aircraft's tire (hydroplaning).

Bearing in mind that a newly paved, ungritted SMA layer is marked by a high degree of slipperiness, some problems with achieving the intended level of friction may develop. On the other hand, gritting, even when followed by sweeping with mechanical brooms to remove unbonded grains, gives rise to a real threat to airplanes due to FOD; thus gritting has generally been ruled out. So the proper level of friction to be obtained by an adequate grain size distribution and the application of coarse aggregate with a high polished stone value (PSV) index.

Requirements for asphalt mixes for airfields mainly comply with specifications for highway engineering in some countries; however in other countries, special technical requirements have been drawn up specifically for airfields. They differ from those for highway applications. But, in each case the ICAO specifications for the finished pavement should be satisfied.

One should also remember that binders applied to specified parts of airfield pavements should be checked for resistance to fuel and deicers because of the destructive actions of these substances. Some manufacturers of road binders offer special products for such applications. Also some research has been done in this area (Steernberg et al., 2000). The following two documents in a series of European Standards established adequate methods of testing:

- EN 12697-41, Asphalt Mixtures—Test Methods for Hot Mix Asphalt—Part 41: Resistance to Deicing Fluids
- EN 12697-43, Asphalt Mixtures—Test Methods for Hot Mix Asphalt—Part 43: Resistance to Fuel

According to EN 13108-20, the aforementioned tests of resistance to deicing fluids and fuels are specifically applicable to SMA used on airfields.

13.1.2 PROS AND CONS OF SMA FOR AIRFIELDS

Despite the fast growth of applications on road pavements, SMA does not enjoy great success on airfields. This is because of the unique requirements for airfield pavements and special problems with SMA, which include the following:

- Combining high macrotexture with low permeability
- Low-initial antiskid properties and the impossibility of using gritting (FOD risk)
- Risk of segregation (e.g., fat spots, porosity) locally changing the surface properties

Regardless of the cited weak points of SMA, airport managers sometimes decide to apply it, taking into consideration the strong points known from the road industry. A comprehensive survey of SMA applications on airports may be found in Prowell et. al. (2009).

13.1.3 AIRPORT IN FRANKFURT ON THE MAIN

One of the best-known applications of an SMA mixtures for airfields is the northern runway of the airport in Frankfurt on the Main (Fraport). It has become famous not only because of the SMA application but also because of the atypical arrangement of the reconstruction of the runway pavement. The 2-year long removal of the worn-out concrete pavement and the laying of a new asphalt structure was completed in 2005.

Fraport is one of the largest airports in the world, with a colossal traffic capacity and an impressive number of takeoffs and landings—more than 200,000 per year. The replacement of the northern runway pavement was carried out exclusively at night to reduce reconstruction-related air traffic problems. This necessitated organizational and technological arrangements to complete each stage of the reconstruction work early in the morning. The intended entrance to the work site was at 10:30 p.m. on a given night, followed by the first landing at 6:00 a.m. next day. The scope of work was divided into independent day lots (actually night lots), which had to be executed within a 7.5-hour break in takeoffs and landings.

The runway intended for rebuilding was 4000 m long and 60 m wide, for a total of 240,000 m². The working lot reconstructed over one night was 15 m long. It took 300 nights to accomplish the pavement replacement. Work was started near the end of April 2003 and finished in June 2005. The old structure of the runway pavement comprised a set of cement concrete layers. All the new asphalt layers contained polymer modified bitumen (PmB) 25 or PmB 45, according to the German guidelines for modified binders TL PmB 01. They were additionally supplemented with Fisher–Tropsch wax (Sasobit) to lower the compaction temperature of the asphalt. The Sasobit additive made possible the compaction of the asphalt mix at a lower temperature (usually at approximately 130°C) and faster cooling of the wearing course (the temperature after laying and compaction was about 100°C); this was critical since the mat had to cool within 1.5 hours prior to the first landing or takeoff at 6 a.m. The SMA 0/11 wearing course was 40 mm thick. Limestone with a hydrated lime additive was used as a filler. The SMA wearing course was spread at a speed of 5 m/min. Seven rollers were engaged in compacting the mat. At the moment of the first landing, the temperature of the pavement ranged between 52 and 79°C (Sasol GmbH, undated).

The replacement of Fraport's northern runway pavement was successfully completed. At the same time it became one of the best-known runway reconstruction projects carried out without the disruption of transport services.

13.1.4 GARDERMOEN AIRPORT IN OSLO

During the 1990s, a few airfield pavements paved with SMA mixtures were constructed in Norway. Various binders and additives were used, depending on the climatic zones in which they were placed. The biggest of these airfields with an SMA pavement is Gardermoen near Oslo. On Gardermoen's runways, 4-cm thick SMA mix of 0/11 mm were placed on the western runway) and 4-cm thick SMA mix 0/16 mm was placed on the eastern runway. Two SMA runways, 3300 m and 2950 m long, were constructed there with styrene-butadiene-styrene (SBS) modified binders (Larsen, 2002).

13.1.5 JOHANNESBURG AIRPORT

In 1999 the following comparative trial sections of different mixes applied in new asphalt wearing courses were laid down at the international airport in Johannesburg, South Africa (Joubert et al., 2004):

- 0/19 mm asphalt concrete (continuous graded coarse mix)
- 0/9 mm SMA with tested parameters: binder content 7.1% (40/50 Pen type), voids in mix 5.7%, stability 6.7 kN, flow 4.0 mm, passing by sieve 2.36 mm 17%, density in place 92%
- 0/13 mm porous asphalt concrete

These mixes were also compared with the existing old wearing course of continuously graded mix.

All sections were located in the landing area, a zone of heavy dynamic loads. Tests were aimed at determining the practicality of various mixes, assuming ungrooved pavements. The surface integrity of the section of pavement and its surface properties were inspected periodically. Special attention was paid to antiskid properties, lifespan, and the buildup of rubber with time (worn-off the airplane tires). The results for the SMA section were as follows:

- Grip number—initially 0.64, after 5 months 0.71
- Surface texture—initially 1.33 mm, after 5 months 0.9 mm

The summary of SMA's performance on the trial section pavement proved that the surface properties of the tested SMA layer were better than the other mixes tested in the trial sections. The porous mix was also recognized for its good characteristics, with the exception of its durability, which was the lowest of those tested here. The conventional asphalt concrete pavement demonstrated poor antiskid properties and therefore needed grooving (Joubert et al., 2004).

13.2 SMA ON BRIDGE DECK STRUCTURES

Surfacing on bridge structures is not, and should not be, like that of a standard pavement on a soil subgrade. The essential difference lies in a different mode of operation. There are special circumstances that must be considered, including the following:

- The cooling and warming effect developing from underneath the bridge deck pavement caused by changes in air temperature under the steel structure and faster changes of the pavement temperature due to wind action, which occurs faster and more intensely than in case of a pavement on grade
- Structural deflections of a bridge's deck caused by passing vehicles
- The amplitude of bridge deck vibration, which is much higher than that of conventional road pavement
- Much more intensive applications of deicers, leading to the quick degradation of asphalt mixes applied on bridges

For all these reasons, asphalt pavements on bridge decks are subjected to faster deterioration than their soil subgrade equivalents. Therefore, when designing a combination of bridge pavement courses, some additional points have to be observed as follows:

- The critical element influencing the pavement service life is the durable bonding of all the layers together (asphalt courses with a protection layer and the deck).
- The more flexible the structure, the more elastic the asphalt layers should be.
- Good compaction of the layers should be taken into consideration because it results in low-water permeability, although rolling on a low-stiffness bridge is challenging.

The deflections of orthotropic plate structures are usually higher than structures with cement concrete deck slabs. Consequently, when asphalt mixes are constructed on steel orthotropic structures, the most frequently applied asphalt mixes are those with the highest fatigue strengths (e.g., mastic asphalt with a highly modified binder) (Damm and Harders, 2000). In some countries, fine-graded SMA has been also used (see Section 13.2.1). Some interesting concepts and analyses can be found in several papers dealing with this subject (Huurman et al., 2003; Medani, 2001a; Medani, 2001b).

13.2.1 Examples of SMA Pavements on Bridges

What follows are descriptions of a few applications of SMA-type asphalt mixes for pavements on steel bridge decks. It has been noted before that this type of bridge construction poses the greatest challenge for the asphalt pavement.

13.2.1.1   Bridge in Roosteren, the Netherlands

In 2005, in Roosteren, the Netherlands, an experimental SMA pavement was made with mastic containing a binder that was highly modified with elastomer (Pen@25°C = 50/70, SP > 90°C) instead of typical mastic asphalt (von Brochove et al., 2008). Demanding requirements, as follows, were required of the SMA mix to compare it with mastic asphalt:

- Resistance to permanent deformation measured by the triaxial compression method after EN 12697-25
- Cracking resistance by the method based on the semicircular bending test at 0°C and 5°C after EN 12697-44
- Fatigue limit, four-point bending test, prismatic sample (4PB-PR) at 5°C after EN 12697-24
- Stiffness, four-point bending test, prismatic sample (4PB-PR) at 5°C after EN 12697-26

The test results proved that the designed mix had a very high fatigue limit. It was laid down in one layer with gritting performed. The grading of the SMA aggregate mix is shown in Figure 13.1.

13.2.1.2   West Bridge on the Great Belt Link, Denmark

Another example of an SMA bridge application in Europe is found on one of the longest bridges in Europe—the Great Belt Link connecting Denmark with Sweden.

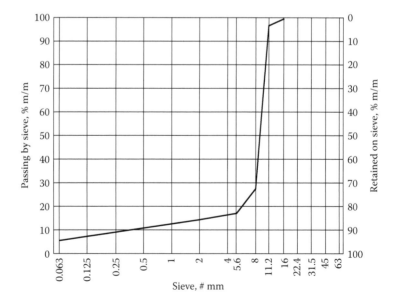

FIGURE 13.1   Grading of the SMA aggregate mix, Roosteren design. (Data from von Brochove, G.G., Voskuilen, J., and Visser, A.F.H.M., *Proceedings of the 4th Eurasphalt & Eurobitume Conference,* Copenhagen, paper 402–102, 2008.)

It is a 6600-m long, prestressed concrete structure with an asphalt pavement consisting of the following courses (Wegan, 2000):

• 15 mm open graded asphalt concrete (drainage layer)
• 40 mm thick asphalt concrete (protective course)
• 40 mm thick SMA (wearing course)

The expected lifespan of the SMA layer on this bridge is 25 years.

13.2.1.3 Bridge in Płock, Poland

The bridge in Płock, Poland, is a steel, highway and railway, multispan bridge with a significant longitudinal slope. During reconstruction in 1998, a new sprayed protection layer was applied and one layer of 0/16 mm SMA surfacing was laid down. An SBS modified binder—with 50/80 Pen@25°C, SP > 63°C, and ER > 80%—was used. It was an experimental SMA application of 0/16 mm grading on a steel bridge. Its surface integrity, after 2 years in operation, was at least a warning. Large areas of cracks and slight rutting could be seen here and there (Figure 13.2.a). The pavement will have to be reconstructed soon. This rapid pavement failure was probably caused by poor adhesion of the course to the protection layer and high permeability resulting from the use of the coarse 0/16 mm grading.

13.2.1.4 Bridge in Wrocław, Poland

The bridge in Wrocław, Poland, is a steel construction with short spans but substantial deflections and vibrations. A completely new pavement was laid in 1997. It consisted of an asphalt mastic protection layer (2 cm) and two 0/8 mm SMA layers—the first one as an intermediate layer and the second one as a wearing course. The mastic layer was spread manually, whereas both SMA courses were

(a)   (b)

FIGURE 13.2 The bridges in Płock and Wrocław: (a) condition of the Płock bridge pavement; (b) condition of the Wrocław bridge pavement. (Photos courtesy of Krzysztof Błażejowski.)

laid mechanically. An SBS modified binder—with 50/80 Pen@25°C, SP > 53°C, and ER > 50%—was used in all the asphalt layers. The condition of the pavement after 13 years in operation was still good, with no cracks or potholes (Figure 13.2.b). Slight rutting was observed (the bridge is located at the approach to a crossing with traffic lights), but repair was needed only in the area of the joints.

13.3 THIN SMA COURSES

Considering the technology of thin SMA layers, but not only gradation of the aggregate blend, but the quantity and type of binder as well, must provide a mix that can be placed by mechanical spreading of a layer up to approximately 4 cm (usually less than 3 cm) thick using a standard paver. Furthermore, the component materials and the final mix itself have to produce a layer that has the following qualities (Sybilski and Styk, 1996):

- Is impervious to water and deicers (excluding porous asphalt)
- Has a suitably high coefficient of friction
- Is resistant to permanent deformation
- Is resistant to low-temperature cracking
- Is resistant to fatigue
- Has the potential to reduce traffic noise

It is also worth remembering that, despite their many strengths, thin courses neither reinforce the pavement substantially nor solve the problem of fatigue (net) and reflected (transverse) cracking (Pandyra et al., 1994). Many countries have their own original technological solutions for courses less than approximately 4 cm thick, which usually consist of various gap-graded mixes, including SMA.

Finer SMA (e.g., 0/5, 0/8) mixes are used for thin layers rather than the 0/11 mm and 0/12.5 mm SMAs. These finer SMAs are noted for many good points, such as similar or only slightly worse resistance to permanent deformation compared with 0/11 mm and 0/12.5 mm mixes. The finer SMAs are also characterized by lower water permeability at the same void contents as coarser SMAs (Cooley and Brown, 2003). Additionally, higher contents of binder in the finer mixes lead to an increase of durability and improved mix workability. If properly designed, finer SMAs tend to reduce the appearance of fat spots, so there is a possibility to reduce the stabilizer content.

The following remarks about SMA technology deserve mention:

- Using modified binders is preferred; excessively hard binders are not recommended.
- Proper bonding between the thin course and its sublayer is one of the most important factors determining the pavement's durability; special tack coats of asphalt emulsions (with enough hard binder) are preferred.
- Usually fewer roller passes are needed for the suitable compaction of a thin course; however, the high speed of laying requires a proper number of rollers to keep up with the paver speed.

- Excessively heavy rollers should not be used because they may crush aggregate grains; vibration can only be used occasionally and with great care; on ultra-thin courses the vibrations should be turned off.
- Attention should be paid to temperature drops in the mix during its spreading because thin layers are very susceptible to fast cooling caused by cool crosswinds or a cold sublayer.
- Almost always, a thin SMA layer can be opened to traffic sooner than can a conventional thickness layer; in the case of ultra-thin layers, owing to their rapid cooling, opening to traffic can be done in as little as 30 minutes after the end of compaction (Carswell, 2002).

Thin SMA layers have been used all over the world. Descriptions from Argentina, the United States, the United Kingdom, Sweden, Poland, and other countries are available in literature (Bolzan, 2002; Carswell, 2002; Carswell, 2004; Cooley and Brown, 2003; Richardson, 1997). An interesting review of the performance (e.g., macrotexture, skid resistance connected to aggregates' PSV, visual condition) of thin layer sections after 15 years can also be found (Nicholls et al., 2008).

Examples of thin SMA layers evaluated after a minimum of 10 years in operation can be the most interesting. One case in point may be the wearing course on the DK3 route in Poland. The 2.5- to 3-cm thin SMA layer of 0/6.3 mm grading was laid in 1993. It was still in very good condition after 13 years in operation (Figure 13.3).

13.4   ATYPICAL SMAs

SMA mixes have performed well in wearing courses, so it was a natural process to test them in the remaining layers of a pavement. As everybody already knows, SMA has proved to be a good material for these places. Therefore, SMA has found its way

FIGURE 13.3   Thin SMA wearing course of 0/6.3 mm grading paved on the DK3 route in Poland, condition in 2006 after 13 years in operation. (Photo courtesy of Krzysztof Błażejowski.)

to intermediate layers. Also research on a special type of SMA for low-noise pavements is in progress in Germany.

Besides classic SMA mixes, the method of designing a strong mineral skeleton has encouraged many road engineers to carry out their own trials on new mixes. One of them is Kjellbase, though it is not a true SMA mix.

The latest atypical SMA applications are colorful mixes.

13.4.1 SMA in an Intermediate Course

In some countries, SMA mixes have been applied in lower layers of the pavement structure. They are usually coarse-graded SMA mixes from 0/16 to 0/22 mm. Such solutions have been tested in the United States and recently in Germany, where they are called *Splittmastixbinder* (SMB) (Gärtner et al., 2009; Schünemann, 2006). Because of the high binder contents, commonly a polymer modified binder with a stabilizer as well, the fatigue properties of the pavement are definitely better than those of conventional asphalt concrete. The application of a strong aggregate skeleton increases the resistance to rutting.

In many cases, SMB 0/16 with a hard modified binder may be better than a conventional asphalt concrete layer. Experimental roadway sections in Bavaria (Germany) on the highway A73 are good examples of such an application (Gärtner et al., 2009). The SMB mix has been designed with target air voids in Marshall specimens between 3.5 and 4.0% (v/v) and a minimum binder content of 5.0%. A hard modified binder (pen@25°C = 10/40, SP > 65°C) has been used. Figure 13.4

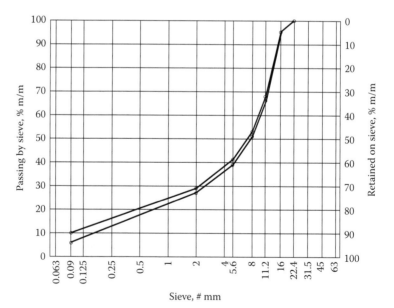

FIGURE 13.4 Grading curves (limits) of the SMB mix used on Highway A73 in Germany. (Data from Gärtner, K., Graf, K., and Schünemann M., Asphaltbinderschichten nach den Splittmastixprinzip. Strasse und Autobahn, July 2009 With permission.)

shows the grading curves of the designed SMB mix. The thickness of the compacted layer has been defined as 7.5 cm, the required compaction factor greater than or equal to 98%, and the content of air voids in the finished course from 2.0 to 6.0% (v/v).

The test results on SMB mixes with hard grade polymer modified binder have confirmed that an intermediate course of this type can be a better solution than the conventional asphalt concrete; with a similar resistance to rutting, the fatigue life of SMB is remarkably better (Schünemann M., 2006).

13.4.2 Low Noise SMA

Earlier, common practice has designated porous mixtures as the most effective way of reducing traffic noise. One- or two-layer pavements have been used, while the latter ones generally are a more effective option. However, that is an expensive solution. On the other hand, it has been found (von Bochove and Hamzah, 2008) that gap-graded mixes composed in accordance with the SMA concept—with an air void content of 9–16% (v/v)—provide a worthwhile alternative to porous asphalt in urban areas. They are marked by a higher resistance to the loads occurring in urban traffic conditions, a longer service life, and good noise reduction properties (up to −5 dB[A]). At the same time, the authors have added that such a mixture cannot be a conventional SMA, but it has to distinguish itself by a significant gap grading and a strong skeleton of coarse particles.

The concept of a "silent" SMA (SMA LA), which is being developed in Germany, is an example of such a solution. Some test sections on roads in Bavaria (Germany) made of SMA LA mixtures 0/5 and 0/8 have been described (Gärtner et al., 2006). The following are the expected values of SMA LA:

- Content of voids above 10% (v/v)
- Gap-graded aggregate mix
- Grading 0/5 mm for layers 15–25 mm thick
- Grading 0/8 mm for layers 20–30 mm thick

The SMA LA course is not gritted since noise reduction has been given high priority. Very positive results of skid resistance with the SKM (SKM – Seitenkraft-Messverfahren – Griffigkeit) (results greater than 0.58) method have been achieved, and noise tests with close-proximity (CPX) method have yielded reductions both at 80 km/h and 120 km/h. Finally the SMA 0/8 LA mixture has turned out to be more effective in noise reduction than the SMA 0/5 LA. Figures 13.5 and 13.6 depict the grading curves of SMA 0/5 LA and SMA 0/8 LA.

Also, in Denmark the road administration, together with industry and consultants, has created a system of classifying the noise reduction effects of various types of asphalt surfacings (Andersen and Thau, 2008). Assessment of the surfacing is carried out according to the CPX method at two speeds, 50 km/h (reference noise level 94.0 dB[A]) and 80 km/h (reference noise level 102 dB[A]). It is worth noting that in Denmark two types of SMA—6 + SRS and SMA 8 SRS—are used for noise-reducing asphalt surfaces as follows:

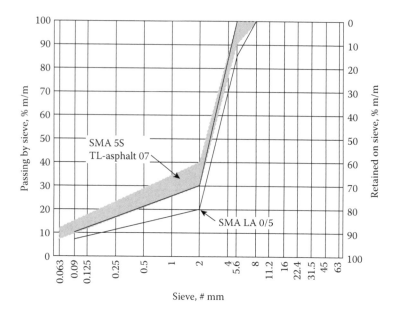

FIGURE 13.5 Comparison of the grading curves of SMA 0/5 LA and typical SMA 0/5S according to TL-Asphalt 07. (Data from Gärtner, K., Graf, K., Meyer, D., and Scheuer, S., Lärmtechnisch optimierte Splittmastixasphaltdeckschichten. Strasse und Autobahn, 12/2006; TL Asphalt-StB 07. Technische Lieferbedingungen für Asphaltmischgut für den Bau von Verkehrsflächenbefestigungen. Ausgabe [in German] 2007.)

- SMA 6 + SRS—maximum aggregate size of 8 mm, air voids between 4 and 10% (v/v), ratio of binder volume to aggregate volume of at least 0.18, minimum thickness of 45 kg/m²
- SMA 6 + SRS—maximum aggregate size of 11 mm, air voids between 3 and 10% (v/v), ratio of binder volume to aggregate volume of at least 0.17, minimum thickness of 50 kg/m²
- SMA 8 SRS—maximum aggregate size of 8 mm, air voids between 4 and 12% (v/v), ratio of binder volume to aggregate volume of at least 0.18, minimum thickness of 55 kg/m²

These requirements are part of the first generation specifications used in tenders (contracts) in Denmark.

13.4.3 KJELLBASE

The strong mineral skeleton applied in SMA has attracted some followers. After all, nothing stands in the way of using similar mixes in intermediate courses or base layers. Kjellbase makes such a mix. The concept, drawn on the "real SMA" idea, came into being in the beginning of 1997, when it was created by Kjell Sardal and S. Gouw. The first trial section was placed in 1997 (Sluer, 2001, 2002; Sluer et al., undated). In fact, Kjellbase is a 0/25 mm gap-graded mix. Now let us have a look at

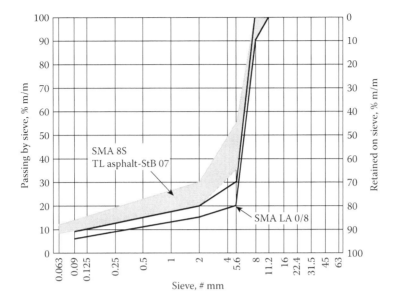

FIGURE 13.6 Comparison of the grading curves of SMA 0/8 LA and SMA 0/8S according to TL-Asphalt 07. (Data from Gärtner, K., Graf, K., Meyer, D., and Scheuer, S., Lärmtechnisch optimierte Splittmastixasphaltdeckschichten. Strasse und Autobahn, 12/2006; TL Asphalt-StB 07. Technische Lieferbedingungen für Asphaltmischgut für den Bau von Verkehrsflächenbefestigungen. Ausgabe [in German] 2007.)

Figure 13.7, which depicts the grading curve of the Kjellbase mix. The gap grading between the 2 mm and 8 mm sieves is clearly visible.

Kjellbase consists of the following (Sluer et al., undated):

- 79% chippings of 8/11 mm, 11/16 mm, and 16/22 mm
- 15% crushed sand
- 6% filler
- 5% modified binder

A small content of fine aggregates (made of only about 5–6% filler and about 15–17% sand fraction) leads to a high void content. Relatively high binder contents close the mix structure, leaving up to 5% (v/v) air voids in the compacted pavement. Since the mix has a large quantity of binder, a mastic stabilizer (e.g., fibers) is required.

Some problems may occur in the laboratory when selecting the right method for the evaluation of mixes of that type. The inventors of the Kjellbase mix concept have estimated that the triaxial compression test may be the proper method.

To summarize the information on Kjellbase, the mineral skeleton of the mix (shown in Figure 2.4.b) is similar to the vision of Figure 2.5. The increased quantity of binder and lower contents of voids in the Kjellbase layer, compared with those of conventional base layers, improves fatigue durability. Thus the mineral skeleton gives that course a higher resistance to permanent deformation.

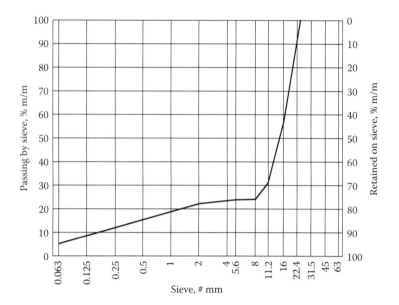

FIGURE 13.7 Kjellbase grading curve. (From Sluer, B.W., Kjellbase. A future without ruts. Development of a heavy duty pavement. Presentation at the 1st International Workshop. SMA and JRS Fibers, Hannover 2002, With permission.)

13.4.4 COLORED SMA

Colored asphalt surfacing is an aesthetically pleasing option. Colored wearing courses can also be executed with fine-graded SMAs (Błażejowski and Styk, 2000) using special synthetic (colorless) binders. Besides their aesthetic appeal, colored surfacing can also be used to mark pedestrian crossings or other safety-related features.

When making colored SMA, it is important to remember to carefully clean the asphalt plant, mixer, and silos, removing "black" mix remains. Loose fibers, specifically those containing no binder additives and bright aggregates, should be applied.

14 European Standards Concerning SMA

A series of European standards—designated EN 13108 and containing requirements for the design, testing, and production of asphalt mixtures—has been implemented in all member states of the European Committee for Standardization (CEN). It consists of 10 standards:

1. EN 13108-1:2006, Bituminous Mixtures—Material Specifications—Part 1: Asphalt Concrete
2. EN 13108-2:2006, Bituminous Mixtures—Material Specifications—Part 2: Asphalt Concrete for Very Thin Layers
3. EN 13108-3:2006, Bituminous Mixtures—Material Specifications—Part 3: Soft Asphalt
4. EN 13108-4:2006, Bituminous Mixtures—Material Specifications—Part 4: Hot Rolled Asphalt
5. EN 13108-5:2006, Bituminous Mixtures—Material Specifications—Part 5: Stone Mastic Asphalt
6. EN 13108-6:2006, Bituminous Mixtures—Material Specifications—Part 6: Mastic Asphalt
7. EN 13108-7:2006, Bituminous Mixtures—Material Specifications—Part 7: Porous Asphalt
8. EN 13108-8:2005, Bituminous Mixtures—Material Specifications—Part 8: Reclaimed Asphalt
9. EN 13108-20:2006, Bituminous Mixtures—Material Specifications—Part 20: Type Testing
10. EN 13108-21:2006, Bituminous Mixtures—Material Specifications—Part 21: Factory Production Control

The standards numbered from 1 to 7 are intended for constructing products harmonized with the Construction Product Directive No. 89/106, which is a *classification* type of standard. This type of standard does not include ready-to-meet sets of requirements, but only a list of properties with a set of categories-levels of requirements. Basically, they are universal standards to be adopted by any CEN-member state.

It is impossible to establish the same requirements for all of Europe due to its substantial climatic differences and diversified road technology experiences. As a result, the common standard has to reflect these differences. The crucial point is that all countries use the same category symbols for SMA properties and the same testing methods for those mixtures. Owing to this *classification* standard, each country may individually specify its essential requirements for asphalt mixtures by placing them in a national application document (NAD). The NAD is understood to be a document introducing an EN standard in national technical regulations, containing a combination of properties and levels of requirements suitable for a given country. An NAD may appear in one of the following form of:

- Technical guidelines (e.g., British PD 6691)
- A national standard that does not contradict the EN standard (e.g., Austrian standard ÖNORM B 3584:2006)
- A National Annex to a national standard implementing the EN standard (e.g., Slovenian Standard SIST 13108-5)

Apart from the standards concerning separate asphalt mixtures (numbered from 1 to 7), the system contains three additional standards (numbered 8, 20, and 21). The standard EN 13108-8 applies to reclaimed asphalt (RAP), therefore it does not refer to a construction product directive. The standard EN 13108-20, entitled Type Testing, concerns the required range of tests necessary for marking a mixture with the CE symbol (for so-called conformity assessment). The last standard, EN 13108-21 Factory Production Control, includes the description of requirements for asphalt plants and quality control. All parts form a complete system of requirements for design, testing, production, and the conformity assessment of asphalt mixtures.

The standard EN 13108-5, which describes the classification system of requirements for SMA mixtures, will be further discussed. Subsequently, the clauses of EN 13108-20 concerning SMA mixtures will be outlined. The present chapter concludes with a description of factory production control according to the standard EN 13108-21.

Any reader interested in details of the presented standards should look at the original texts and clauses.

14.1   THE EUROPEAN STANDARD EN 13108-5

EN 13108-5 defines the requirements for SMA mixtures for use on roads, airfields, and other trafficked areas. This standard should be read along with other standards as follows:

- EN 13108-20:2006, Bituminous Mixtures—Material Specifications— Part 20: Type Testing
- EN 13108-21:2006, Bituminous Mixtures— Material Specifications—Part 21: Factory Production Control
- EN 12697-x, Bituminous Mixtures. Test Methods

The mutual relationships among those standards are displayed in Figure 14.1.

Sets of properties of asphalt mixtures (Figure 14.1) are listed in the standard EN 13108-x (any from 1 to 7). Each of these properties corresponds to an appropriate test method described in a standard from the series EN 12697-x. However, it should be noted that sometimes the standard provides for more than one test procedure for

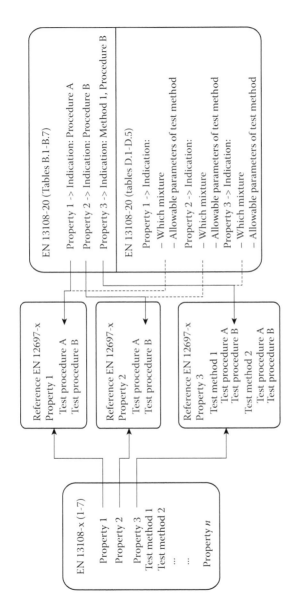

FIGURE 14.1 Example of relationships between European standards concerning asphalt mixtures.

determining properties (Figure 14.1 indicates the procedures A, B, ...). The standard EN 13108-x (1–7) does not specify which procedure to select but indicates that the right method can be found in the standard EN 13108-20, with directly recommended test methods and procedures provided in tables.

14.2 DEFINITIONS

Stone mastic asphalt is defined as a gap-graded asphalt mixture that has bitumen as a binder, and is composed of a coarse crushed aggregate skeleton bound with a mastic mortar.

The standard defines two types of recipes (job mix formulae [JMF]):

- Input target composition—this is the determined composition of the mixture given through listing its constituent materials, the gradation curve, and the percentage content of binder added to the mixture; this formula is the result of laboratory validation of the mixture.
- Output target composition—this is the determined composition of the mixture given through listing its constituent materials, the midpoint gradation, and the percentage of soluble binder content in the mixture,which are obtained as results of the composition analysis (extraction) of a produced mixture; usually this formula is the result of production validation of the mixture.

An additive is defined as a constituent material supplemented to the mixture in small amounts, (e.g., organic or nonorganic fibers and polymers added to enhance mechanical properties, workability, or the color of the mixture).

14.3 IDENTIFICATION OF AN SMA MIXTURE

The standard stipulates that each delivery ticket should furnish at least the following information:

- Name of the manufacturer and mixing plant
- Mix identification code
- Designation of the mixture in the format

        SMA   D   binder

where D is the maximum aggregate size in millimeters and *binder* is the binder type. For example, SMA 11 50/70 denotes an SMA mixture with a gradation up to $D = 11.2$ mm, with the road binder 50/70 (according to EN 12591).

Apart from the aforementioned information, a manufacturer should also provide the following:

- Instructions on how a recipient can obtain detailed information about the compliance of a mixture with the requirements of EN 13108-5

- Details regarding the compliance of a mixture with requirements concerning resistance to fuel and resistance to deicing fluids (if the delivery applies to an airfield)
- Details of additives used

14.4 REQUIREMENTS FOR CONSTITUENT MATERIALS

Only constituents with an established suitability are allowed to be used for SMA mixtures. The established suitability means meeting the material requirements of the following:

- A European standard
- A European technical approval (ETA)
- A documented, positive experience with a specific kind of material in the past in which the reference documents confirm the suitability of the material (e.g., test results combined with observations in places of performance)

14.4.1 BINDERS

Road (paving grade) binder according to EN 12591 and polymer modified binder according to EN 14023 have been used in SMAs. SMA mixtures containing chemically modified binders, which are not classified by the standard EN 14023, are not covered by the clauses of the standard EN 13108-5. Natural asphalts according to EN 13108-4:2005, Annex B, may be added.

14.4.1.1 Road Binder

Road binders after EN 12591 are categorized on the basis of penetration at 25°C and have been incorporated in SMAs. When using them, a binder from the series 30/45 to 330/430 should be selected. This is a very wide range of bitumens (from hard to very soft ones) intended to allow for a wide variety of possible applications depending on local conditions (e.g., climate, traffic loads). In reality, it is for the most part a choice between 50/70 and 70/100.

14.4.1.2 Polymer Modified Binder

Polymer modified binders (PMBs)—which, according to EN 14023, is divided on the basis of penetration at 25°C and softening point (ring and ball)—have also been used.

When a modified binder is being used to enhance a particular property of an SMA mixture, with no reference to binder features (e.g., resistance to fatigue), some additional tests should be performed to confirm the desired effectiveness of a given binder. These tests should be conducted using methods described in the standard EN 12697. Using the results of previous tests is permissible. The origins of this clause of the SMA standard (EN 13108-5) are in the structure of the standard concerning PMB (EN 14023), which has a combination of classes enabling the description of basic requirements for PMB. There is no direct correlation between these requirements and the functional properties of asphalt mixes. Consequently, the effectiveness of a selected PMB in the asphalt mixture should be checked.

14.4.1.3 Natural Asphalts

Natural asphalt may be employed for SMA as an additive to the road binder or modified binder under the following conditions:

- If it conforms to the requirements of EN 13108-4, Annex B, Tables B.1 and B.2, for natural asphalts with high or low-ash contents, respectively
- When natural asphalt is being incorporated in the road or modified binder by means of the following:
  - Intermixing with heated binder in a liquid state in a tank
  - Direct batching into a pugmill in the case of natural asphalt in the form of a powder or granulate with particles not exceeding 10 mm

14.4.2 AGGREGATES

All types of applied aggregate (coarse, fine, all-in,* added filler) should comply with the requirements of EN 13043 selected for a specific use. The appropriate NAD with requirements for SMA aggregates corresponding to the standard EN 13043 should be selected (examples of such requirements are detailed in Chapter 5).

The amount of added filler should be fixed. Hydrated lime and cement may also be used as fillers.

14.4.3 RECLAIMED ASPHALT

According to the standard, the use of RAP for SMA is permissible. The types, quantities, and requirements for RAP to be used in SMA mixes should be specified in an NAD appropriate to the intended use.

RAP should be classified according to EN 13108-8 and should conform to the relevant requirements for a particular application.

The maximum size of a particle in RAP cannot be larger than size D of the SMA mix. The quality of aggregate in RAP cannot deviate from the requirements for a new aggregate to be used in a given SMA.

When the used RAP contains road binder (unmodified) and when the binder added to the mixture is road binder then additionally one of two values should be determined: either the penetration at 25°C or the softening point (R&B) of a mixture created by combining the recovered binder from the RAP with the new binder added during SMA production. The test result (Pen25 or SP) of this binder mixture should meet the requirements for the target (design) road binder selected for a given SMA. Formulae for calculating properties of binder mixes can be found in the standard EN 13108-5, Annex A. The method of recovering binder from recycled asphalt paving mixture should be in accordance with EN 12697-3 (binder recovery—rotary evaporator) or EN 12697-4 (binder recovery—fractionating

---

* According to the definition in EN 13043, all-in aggregate is a granular material consisting of a mixture of coarse and fine aggregates, can be produced without separating into coarse and fine fractions, or can be produced by combining coarse and fine aggregate.

column). The penetration should be determined according to EN 1426 and the softening point according to EN 1427. Such an additional requirement is used in following cases:

- In a wearing course when more than 10% (m/m) RAP is used
- In a regulating (leveling) and intermediate course when more than 20% (m/m) RAP is used

When used RAP or the new SMA contains a modified binder or a modifier additive, according to Item 5.1 of the standard the amount of RAP cannot exceed the following:

- 10% by mass of the total mixture if the SMA is intended for a wearing course
- 20% by mass of the total mixture if SMA is meant for a regulating* or intermediate course or a base course

Both the client and the producer of the SMA mix may arrange otherwise, provided that other local (national) regulations are not infringed upon.

14.4.4 ADDITIVES

Additives should conform to the requirements for constituent materials; namely, they should have the determined suitability—marked by complying with an appropriate EN standard, ETA, or a demonstrable history of satisfactory use.

14.5 REQUIREMENTS FOR AN SMA MIX

An SMA formula has to be documented and demonstrated (declared). Any SMA mixture made according to the recipe has to meet the standard requirements determined by a given country.

14.5.1 GRADATION

The fundamental rules regarding the mix design include the following:

- The gradation should be expressed in mass percentages of the total aggregate mix; the accuracy of percentages passing
  - all sieves (with the exception of the 0.063 mm sieve) should be expressed to 1%.
  - the 0.063 mm sieve should be expressed to 0.1%.
- The content of binder and additives should be expressed in mass percentages of the asphalt mixture, with an accuracy of 0.1%.

---

* Leveling course in the United States.

- The type of fine aggregate used and the adopted ratios in the case of a mix may be given in a recipe or specification.
- The gradation may be described with either "basic sieve set plus set 1*" or "basic sieve set plus set 2[†]"; a combination of sieves from set 1 to set 2 is not permissible.[‡]

The gradation of an SMA mixture should be established with a minimum of five sieves: 0.063, 2.0, D, 1.4D, and the characteristic coarse sieve (a selected sieve between 2.0 mm and D). Basically, the gradation limits, which are given in the standard, must adhere to the rules for preparing NADs to the standard EN 13108-5. Each country, by its NAD, may determine an SMA mix's gradation envelopes, guided by the following:

1. Overall limits on the target composition displayed in Tables 1 and 2 of the standard
2. Permissible ranges between maximum and minimum values on selected sieves (in Table 3)

The standard allows for the use of additional control sieves, called optional (characteristic) sieves, to enable a more precise description of the gradation as follows:

- A characteristic sieve for fine aggregate may be selected between the 2.0 mm and the 0.063 mm sieves; in addition, the standard stipulates the set of sieves to be chosen from 0.125, 0.25, 0.5, and 1.0 mm;
- An optional characteristic sieve for the coarse aggregate may be selected to provide one more additional sieve with a size between 2.0 mm and D.

Finally, to describe the gradation envelope, one can use the following set of sieves:

- 0.063 mm sieve (obligatory)
- Characteristic sieve for the fine aggregate (optional)—sieve between 0.063 and 2.0 mm,
- 2.0 mm sieve (required)
- Characteristic coarse sieve (required)—a selected sieve between 2.0 mm and D
- Additional characteristic coarse sieve (optional)—a selected sieve between 2.0 mm and D
- Sieve D (required)
- Sieve 1.4D (required)

---

* Set "+1": 1.0, 2.0, 4.0, 5.6, 8.0, 11.2, 16.0, 22.4, 31.5, 45.0, 63.0 mm.
[†] Set "+2": 1.0, 2.0, 4.0, 6.3, 8.0, 10.0, 12.5, 14.0, 16.0, 20.0, 31.5, 40.0, 63.0 mm.
[‡] The sieve systems: basic, +1, and +2 are established in EN 13043.

It is worth noting that the freedom to select characteristic sieves gives a chance to choose those sieves that will provide the best possible control of an SMA mixture (e.g., breakpoint sieves). Figures 14.2 through 14.5 depict the position of boundary points for example mixtures SMA 8 and SMA 11 for sieve set + 1 and SMA 10 and SMA 14 for sieve set + 2. As can be seen, the scope of available solutions (positions of overall limits to a target composition) for any of the mixtures is quite broad. Additionally, in the same figures, the German (for SMA 8 and 11) and British (for SMA 10 and 14) gradation envelopes are presented as examples.

14.5.2 BINDER CONTENT

A series of categories of minimum binder contents in SMA mixes, denoted B_{min}, is detailed in the standard. However, maximum binder contents are not defined; therefore an appropriate category of B_{min} should be matched with each SMA specification.

The categories given in the standard have been adopted for a reference density of an aggregate mix equal to 2.650 Mg/m³. For aggregate mixtures with other densities, the required lower limit of the binder content should be modified using the factor α

$$\alpha = \frac{2.650}{\rho_a}$$

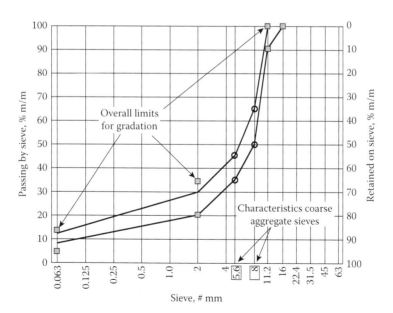

FIGURE 14.2 The position of boundary points of SMA 11S gradation envelopes according to EN 13108-5 and German final gradation limits for this mixture according to TL Asphalt-StB 07.

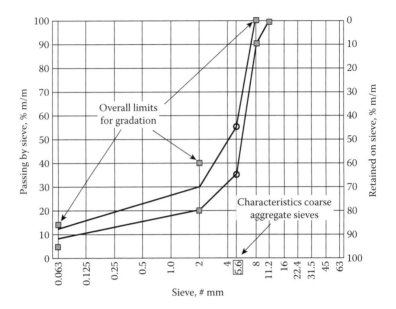

FIGURE 14.3   The position of boundary points of SMA 8S gradation envelopes according to EN 13108-5 and German final gradation limits for this mixture according to TL Asphalt-StB 07.

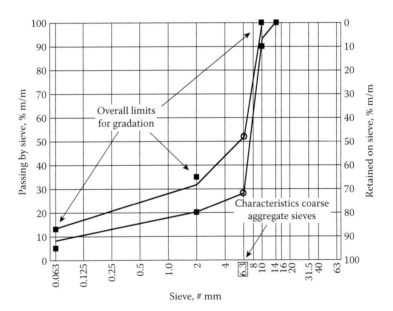

FIGURE 14.4   The position of overall limits to a target composition of SMA 10 according to EN 13108-5 and British final gradation limits for this mixture according to BS PD 6691:2007.

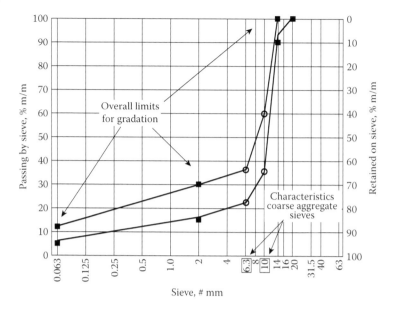

FIGURE 14.5 The position of overall limits to a target composition of SMA 14 according to EN 13108-5 and British final gradation limits for this mixture according to BS PD 6691:2007.

where ρ_a is the particle density of the aggregate mix in megagrams per cubic meter, according to EN 1097-6.

The binder content determined for an SMA mixture should be the sum of all possible sources of binder in it, which includes the following:

- Added binder
- Binder from the RAP (when added)
- Natural asphalt (when added)

Available categories of the minimum binder content in SMA (to be chosen in an NAD or a contract specification) range from 5.0 to 7.6% in 0.2% increments (i.e., $B_{min} = 5.0, 5.2, 5.4, 5.6 \ldots$).*

14.5.3 HOMOGENEITY AND QUALITY OF COATING WITH BINDER

The SMA mixture should be homogenous and completely coated with binder when discharged from the pugmill. There should be no evidence of agglomeration (balling) of the fine aggregate.

---

* Full range: $B_{min}5.0, B_{min}5.2, B_{min}5.4, B_{min}5.6, B_{min}5.8, B_{min}6.0, B_{min}6.2, B_{min}6.4, B_{min}6.6, B_{min}6.8, B_{min}7.0, B_{min}7.2, B_{min}7.4, B_{min}7.6.$

14.5.4　Void Content

Void content is one of the essential properties of a compacted SMA mix. Much attention was paid to this subject in the previous chapters, particularly in Chapters 6 and 7. Establishing the method of sample preparation and suitable conditions (compactive effort and compaction temperature) is necessary for the proper determination of air voids in compacted samples. The standard categorizes of minimum and maximum void contents in SMA samples to be selected in an NAD are as follows:

- Minimum void content in SMA samples range from 1.5 to 6% in increments of 0.5% (i.e., $V_{min} = 1.5, 2. 2.5...$)* plus $V_{min}NR$, where $V_{min}NR$ means no requirement.
- Maximum void contents in SMA samples range between 3 and 8% in 0.5% increments (i.e., $V_{max} = 3, 3.5, 4,...$)† plus $V_{max}NR$, where $V_{max}NR$ means no requirement.

14.5.4.1　Preparation of Samples

The method of preparing SMA samples in the laboratory to determine the void content is specified in the standard EN 13108-20, Item 6.5, with details in Annex C (Table C.1). The NAD should provide values of compactive efforts. Permissible methods include the following:

- Impact compaction according to EN 12697-30, with possible energies or 2 × 25 blows, 2 × 50 blows, 2 × 75 blows, or 2 × 100 blows
- Gyratory compactor according to EN 12697-31, with different numbers of gyrations

The standard EN 13108-20 also states that the JMF should clearly state the adopted method and prevailing conditions of the sample preparation.

14.5.4.2　Determination of the Void Content

The determination of the void content in compacted samples should be evaluated according to the standard EN 13108-20, Table D.2, as follows:

- Bulk density of a sample should be determined according to EN 12697-6, Procedure B (Saturated Surface Dry [SSD]).
- Maximum density of sample should be determined according to EN 12697-5, Procedure A (with the use of water).
- Calculating the void contents in compacted samples should be conducted according to EN 12697-8 (based on formulae given there).

---

* Full range: $V_{min}1.5$, $V_{min}2$; $V_{min}2.5$, $V_{min}3$, $V_{min}3.5$, $V_{min}4$, $V_{min}4.5$, $V_{min}5$, $V_{min}5.5$, $V_{min}6$, $V_{min}NR$.
† Full range: $V_{max}3$, $V_{max}3.5$, $V_{max}4$, $V_{max}4.5$, $V_{max}5$, $V_{max}5.5$, $V_{max}6$, $V_{max}6.5$, $V_{max}7$, $V_{max}7.5$, $V_{max}8$, $V_{max}NR$.

If determining the void contents in a gyratory compactor at a set value of gyrations is required, testing should be conducted according EN 12697-31. In this case, methods of direct measurements of density should not be employed.

14.5.5 Voids Filled with Binder

The standard categorizes minimum and maximum percentages of voids filled with binder (VFB). The following are the available categories of requirements and their denotations:

- Minimum percentage of VFB range from 71 to 86% in increments of 3% (i.e., $VFB_{min} = 71, 74, 77....)^*$ plus $VFB_{min}NR$, where $VFB_{min}NR$ means no requirement.
- Maximum percentage of VFB range from 77 to 92% in increments of 3% (i.e., $VFB_{max} = 77, 80, 83...)^+$ plus $VFB_{max}NR$, where $VFB_{max}NR$ means no requirement.

14.5.6 Binder Draindown

Tests of binder draindown should be performed according to EN 12697-18. Available categories of the maximum permitted binder draindown from the SMA mixtures are $D_{0.3}$, $D_{0.6}$, $D_{1.0}$, and D_{NR}, where D_{NR} means no requirement. However, EN 13108-20 does not indicate directly which method of EN 12697-18 should be used—basket or Schellenberg (see Chapter 8).

14.5.7 Water Sensitivity

Water sensitivity, denoted as indirect tensile strength ratio (ITSR), is determined according to EN 13108-20, Clause D.3. Water sensitivity is determined by test method after EN 12697-12 at a test temperature of 15°C. The available categories of requirements ITSR are $ITSR_{90}$, $ITSR_{80}$, $ITSR_{70}$, $ITSR_{60}$, and $ITSR_{NR}$, where $ITSR_{NR}$ means no requirement.

14.5.8 Resistance to Abrasion by Studded Tires

Resistance to abrasion by studded tires is determined according to EN 13108-20, Clause D.4, and testing after EN 12697-16, Procedure A. The available categories of requirements are Abr_{A20}, Abr_{A24}, Abr_{A28}, Abr_{A32}, Abr_{A36}, Abr_{A40}, Abr_{A45}, Abr_{A50}, Abr_{A55}, Abr_{A60}, and Abr_{NR}, where Abr_{NR} means no requirement.

---

* Full range: $VFB_{min}71$, $VFB_{min}74$, $VFB_{min}77$, $VFB_{min}80$, $VFB_{min}83$, $VFB_{min}86$, $VFB_{min}NR$.
+ Full range: $VFB_{max}77$, $VFB_{max}80$, $VFB_{max}83$, $VFB_{max}86$, $VFB_{max}89$, $VFB_{max}92$, $VFB_{max}NR$.

TABLE 14.1

Test Methods of SMA Resistance to Permanent Deformation and Test Parameters

Device	Method	Test Temperature (°C)	Number of Cycles[a]
Small device,	In air	45	10,000
Method B		50	10,000
		60	10,000
Large device	In air	50	30,000
		60	10,000
		60	30,000

Source: Modified from EN 13108-20. Bituminous Mixtures—Material Specifications—Part 20: Type Testing Table D.1 of Clause D.6.

[a] 1 cycle = 2 passes of wheel.

14.5.9   Resistance to Permanent Deformation

Resistance to permanent deformation is one of the most significant properties. Testing is carried out according to the standard EN 12697-22. The equipment used for SMA testing includes a large size device and a small device.

14.5.9.1   Selection of Device and Test Parameters according to EN 13108-20

For SMA tests according to EN 13108-20, methods listed in Clause D.6 are shown in Table 14.1. The appropriate method adopted in an NAD with appropriate test parameters should be selected from these methods.

The selection of small and large devices is based on Table B.5 of the standard EN 13108-20, which states the following:

- The small device is for testing SMA mixtures designed for axle loads less than 13 tons.
- The large device is for testing SMA mixtures designed for axle loads equal to 13 tons and more.

14.5.9.2   Categories of Requirements according to EN 13108-5

There are three tables (12, 13, and 14) in the standard EN 13108-5 with categories of requirements for resistance to permanent deformation. They are used for the testing method discussed in Section 14.5.9.1. The outline of categories of requirements contained in Tables 12, 13, and 14 of the standard EN 13108-5 are as follows:

- Table 12—for results from the large device; categorizes the following requirements marked with the symbol P as the maximum proportional rut depth (percent): P_5, $P_{7.5}$, P_{10}, P_{15}, P_{20}, and P_{NR}, where P_{NR} means no requirement.

- Table 13—for results from the small device; Procedure B, testing in air, categorizes the following requirements marked with the symbol WTS_{AIR} as the maximum wheel tracking slope (mm/1000 load cycles): WTS_{AIR} 0.03, WTS_{AIR} 0.05, WTS_{AIR} 0.07, WTS_{AIR} 0.10, WTS_{AIR} 0.15, WTS_{AIR} 0.30, WTS_{AIR} 0.40, WTS_{AIR} 0.50, WTS_{AIR} 0.60, WTS_{AIR} 0.80, WTS_{AIR} 1.00, and WTS_{AIR} NR, where WTS_{AIR} NR means no requirement.
- Table 14—for results from the small device; Procedure B, testing in air, categorizes the following requirements marked with the symbol PRD_{AIR} as the maximum proportional rut depth (percent of slab's thickness): PRD_{AIR} 1.0, PRD_{AIR} 1.5, PRD_{AIR} 2.0, PRD_{AIR} 3.0, PRD_{AIR} 4.0, PRD_{AIR} 5.0, and PRD_{AIR} NR, where PRD_{AIR} means no requirement.

14.5.9.3 Additional Conditions for Preparing Samples

The SMA wheel tracking test is carried out on samples (slabs) prepared in a laboratory or cut out of a pavement. Additional requirements for samples may be determined in the following two ways:

- Requiring a void content in the slab; in this case the requirement has a range of 3% (v/v), (e.g., 3–6% [v/v]).
- Requiring the compaction factor; in this instance a requirement with a range of 2% (e.g., 98–100%) should be imposed.

All the aforementioned requirements actually specify the quality of the prepared sample (slab) before testing.

14.5.10 REACTION TO FIRE

When an SMA mix has to meet requirements for resistance against fire specified in other regulations, this property should be tested and classified according to the method described in EN 13501-1.

14.5.11 RESISTANCE TO FUEL ON AIRFIELDS

This requirement regarding resistance to fuel exclusively applies to SMA mixes for airfields. The method of testing this property of SMA is elaborated in the standard EN 12697-43. The categories of the requirements are good, moderate, poor, or NR (no requirement).

14.5.12 RESISTANCE TO DEICING FLUIDS FOR APPLICATION ON AIRFIELDS

This requirement concerning resistance to deicing fluids applies to SMA mixes for airfields only. The method of testing this property of SMA is described in the standard EN 12697-41. The test is conducted using the pull-off method on samples subjected to conditioning in a deicing fluid and on nonconditioned, comparable samples. The categories of requirements are marked with the indexed symbol β: β_{100}, β_{85}, β_{70}, β_{55}, and β_{NR}, where β_{NR} means no requirement. The β is a ratio of conditioned to unconditioned sample result.

14.6 TEMPERATURES OF ASPHALT MIXTURES

The recommended temperatures of SMA mixtures with road bitumens according to EN 12591 are given in the standard EN 13108-5. The minimum temperature of delivery to a laydown site and the maximum production temperature in an asphalt plant are provided there. When using a modified or special binder, one should follow the information passed on by its producer or as determined in other documents.

Example temperatures for two of the most common road bitumens (EN 12591) used for SMA are as follows:

- 50/70— minimum temperature 150°C, maximum temperature 190°C
- 70/100— minimum temperature 140°C, maximum temperature 180°C

14.7 EVALUATION OF CONFORMITY

Asphalt mixtures are construction products and the standards EN 13108-1 to 13108-7 are harmonized with the Construction Products Directive 89/106/EEC. The system "2+" has been adopted to evaluate conformity of asphalt mixtures. It consists of the following:

- Initial type testing of each mix produced by its manufacturer
- Certification of a Factory Production Control (FPC) with reference to EN 13108-21 conducted by a notified body, separate for each production site (asphalt plant)

The SMA design process (recipe) should be followed by confirmation that the mixture meets all requirements shown in the relevant categories listed in the NAD; if it does meet the requirements, it cannot enter the market. This set of tests, called initial type testing, is based on requirements of the standard EN 13108-20 (Table B.5), which include the following:

- Binder content EN 12697-1 and 12697-39
- Grading EN 12697-2
- Void content including VFB EN 12697-8
- Void content of gyratory compacted samples EN 12697-31
- Binder drainage EN 12697-18
- Water sensitivity EN 12697-12
- Resistance to abrasion by studded tires EN 12697-16
- Resistance to permanent deformation EN 12697-22
- Resistance to fuel (airfields) EN 12697-43
- Resistance to deicing fluids (airfields) EN 12697-41

Routine (daily) testing of a manufactured mix is conducted using a system of FPC according to the standard EN 13108-21.

14.8   FACTORY PRODUCTION CONTROL EN 13108-21

In the European system of standardization of asphalt mixtures according to the series of standards EN 13108, the issues of control and quality assurance have been deliberated in EN 13108-21:2006 Bituminous Mixtures—Material Specifications— Part 21: Factory Production Control (with Amendment AC/September 2008). This standard contains an outline of requirements regarding the system of quality assurance during the production process of a mix. Generally speaking, the FPC is in conformity with principles from ISO 9001, so that part of the standard will not be discussed here. Requirements for the production control can be found in the normative Annex A, entitled "Tolerances and Test Frequencies for Finished Asphalt," which stipulates admissible deviations and frequencies of controlling manufactured and delivered mixes.

14.8.1   LEVELS OF REQUIREMENTS

EN 13108-21 provides three levels of requirements related to the expected accuracy of production—level X, level Y, and level Z.

Level Z is a basic one and its application secures the fulfillment of the requirements connected with the evaluation of conformity. Levels X and Y require assurance of a higher frequency of control testing and can be adopted when such increased control is required. The determination of different levels of requirements for various types of mixes or specified contracts is also possible (e.g., a higher level for bridge deck surfacing). In these circumstances, the selection of an appropriate level (X, Y, or Z) indicates the significance of the minimum risk of nonconformity. Consequently, the frequency of testing production samples depends on the accepted level of requirements (i.e., the desired level of conformity).

14.8.2   GENERAL GUIDELINES FOR A CONTROL SYSTEM OF PRODUCTION HOMOGENEITY

The guidelines for controlling production homogeneity are made up of the following two parts:

- Control of compliance with a recipe, accomplished by establishing an operation compliance level (OCL), which is a variable index that estimates the production accuracy and indicates the appropriate mixture test frequencies
- Control of production variability through determining a running mean of the deviation from target (i.e., from a recipe).

14.8.3   CONTROL OF COMPLIANCE WITH AN SMA MIX RECIPE

14.8.3.1   Determination of the Operating Compliance Level

The principle of fixing frequencies of control testing depends on the number of deviations found and has been adopted in the standard EN 13108-21. With an established level of requirements (X, Y, or Z), the frequency of sampling is variable, depending on the production accuracy of the asphalt plant (i.e., the number of deviations from a recipe).

14.8.3.2   Start in Control

The initial frequency of sampling depends on an established category of conformity (X ,Y, or Z), so at first the frequency of sampling is as follows:

- Every 150 tons of a manufactured mixture (for level X)
- Every 250 tons of a manufactured mixture (for level Y)
- Every 500 tons of a manufactured mixture (for level Z)

Samples should be representative of the entire production; the relevant clauses of EN 12697-27 and EN 12697-28 should be taken into account.

14.8.3.3   Analysis of Mixture Composition

Each sample is subjected to extraction testing, resulting in a gradation of the aggregate mix and a content of soluble binder. The sieving operation should be conducted using a set of sieves as in Table 14.2. The standard stipulates testing the aggregate gradation through five sieves (plus possibly a sieve for oversize particles 1.4D). The small number of test sieves speeds up the control analysis but at the same time increases the responsibility of the mix producer to achieve other final SMA properties determined by a recipe. It appears that the proper selection of optional sieves for the coarse aggregates that control the skeleton is of great significance.

14.8.3.4   Selection of a Method for Estimation of Extraction Results

Prior to the comparison of extraction results with admissible deviations, the method for assessment of extraction results should be selected. The choice should be made between the following two methods:

- Single result method—test results of individual samples are treated independently and assessed in comparison with admissible deviations from a target in accordance with Table A.1 of the standard; the sets of the previous 32 results form the basis for the OCL assessment (classified as conforming or nonconforming).
- Mean of four results method—test results of individual samples are treated in groups of four analyses of the same mixture, and the mean result of each group is compared with admissible deviations from a target (classified as conforming or nonconforming); after a comparison of results of the eight mean results (e.g., eight groups, each with four results for a total of 32), they are classified as conforming or nonconforming and form the basis for the OCL assessment.

Thus in the single result method, we assess 32 individual results; in the mean of four results method, we assess eight results (means) calculated for groups of four single results.

TABLE 14.2
Comments on Requirements for Permitted Deviations from the Target for Mixtures Produced Using EN 13108-21, for Small Aggregate Mixtures with D <16mm—the Single Result Method

Percentage Passing (% m/m)	Small Aggregate Mixes (D <16 mm)		Comment
	Individual Samples Tolerance about Target Composition (Single Result Method for OCL Calculation)	Permitted Mean Deviation from Target (Control of Production Variability)	
1.4 D	—	—	The sieve 1.4D—additional sieve on which requirements for passing 98–100% have been established
D	−8/ + 5	±4	The sieve determining maximum particle size in a mix (without oversize particles), (e.g., the record SMA 11 means D = 11.2 mm [at the same time the boundary sieve for oversize particles])
D/2 or characteristic coarse sieve	±7	±4	The sieve indicated in an NAD for the standard EN 13108-5 for each SMA mixture
2.0 mm	±6	±3	—
Characteristic fine sieve	±4	±2	The characteristic sieve has not been determined in EN 13108-21; it should be established in an NAD for EN 13108-5 for each SMA mixture; the choice has been limited to one of the sieves: 0.125, 0.25, 0.5, or 1.0 mm
0.063 mm	±2	±1	Sieve of filler fraction
Soluble binder content	±0.5	±0.3	Soluble binder content with extraction after EN 12697-1

Source: Modified from EN 13108-21. Bituminous Mixtures—Material specifications—Part 21: Factory Production Control.
Note: NAD = National application document; SMA = stone matrix asphalt.

These two methods may not be applied in parallel. In other words, only one method can be used at the same time in the same asphalt plant.

Analyses of deviations using only the single result method will be discussed later.

14.8.3.5 Determination of Compliance of a Tested Sample with a Recipe

After extraction, the result of each sample is compared with admissible deviations on test sieves and soluble binder content. A mix may be regarded as compatible with a recipe when all its components are within admissible tolerances.

Table 14.2 depicts an excerpt of Table A.1 of Annex A of the standard demonstrating admissible deviations from a target appropriate for producing SMA mixes with D less than 16mm (e.g., the group of small aggregate mixtures using the single result method). Using admissible deviations, the accuracy of measuring methods has already been taken into account.

14.8.3.6 Reaching the OCL Assessment

After assessing the mixture extraction results (the set of 32 results), one can determine the OCL level. According to the standard, there are three OCL level—A, B, and C. They can be roughly interpreted as follows:

- OCL A proves a high rate of compliance with the target for manufactured mixtures.
- OCL B is an intermediate assessment and is still good.
- OCL C is a warning assessment for asphalt plant personnel (and their customers) that the composition of produced mixtures fluctuates considerably.

The OCL assessment is established after 32 consecutive analyses, but it should be kept in mind that they do not have to be samples of only one mixture type; the standard stipulates taking into consideration 32 consecutive results of any produced mixture. The selection of an appropriate OCL then follows, considering the total number of samples that do not conform with requirements

Number of nonconformances	OCL
2 or less	Level A
3–6	Level B
7 or more	Level C

In the event that more than 8 of the latest 32 results are nonconforming, the plant's equipment and procedures should be subject to an immediate and thorough review.

The OCL assessment is a dynamic process because the latest 32 results are always used for calculations. After startup at the beginning of the construction season, the test frequency at the starting level should correspond to the lowest OCL-C of a designated level of requirements (X, Y, or Z). Sampling frequency is subject to change weekly based on the lowest OCL achieved during the previous calendar week.

Minimum frequencies of testing samples in relation to the achieved OCL from the previous week are valid for the next calendar week. The following are OCL sample testing frequencies for the single result method:

OCL-A	Level X: 600 tons	Level Y: 1000 tons	Level Z: 2000 tons
OCL-B	Level X: 300 tons	Level Y: 500 tons	Level Z: 1000 tons
OCL-C	Level X: 150 tons	Level Y: 250 tons	Level Z: 500 tons

Interestingly, the better the OCL, the lower the test frequency. These data also show that for each category (X, Y, or Z), when the OCL advances by one level, the test frequency decreases by half (or twice as much material can be produced between tests).

The startup of a new plant means an operational startup with the OCL-C assessment. A shutdown of longer than 3 months during the season or a major repair of the asphalt plant results in the OCL being reduced by one level from where it was before the stoppage. The OCL level cannot be changed until 32 results have been obtained and the first new OCL assessment is available. In the case of an asphalt plant with a low rate of production, the standard stipulates testing the composition of a mix at least every 5 operating days. Mobile coating plants, after relocation, are treated like plants shut down for 3 months or restarted after a major repair.

14.8.4 CONTROL OF PRODUCTION VARIABILITY

The standard EN 13108-21 contains a second element of FPC—namely, the control of production variability. This control is applied through the determination of the running mean deviation from the target and is described in Item A.5 of the standard. It involves monitoring systematic trends at the production stages of asphalt mixtures to prevent permanent one-sided deviations. The following parameters of asphalt mixtures are under such supervision:

- Mass percentage of material passing through the sieve D
- Mass percentage of material passing through the sieve D/2 or the characteristic coarse sieve
- Mass percentage of material passing through the 2 mm sieve
- Mass percentage of material passing through the 0.063 mm sieve
- Soluble binder content

Calculations of the mean value of deviations from the target should be conducted for each of these properties. The mean deviations calculated on an ongoing basis should be compared with admissible values given in Table A.5 of the standard (Table 14.2.).

Running mean analyses are undertaken for the following groups of mixes (clause A.2 of standard):

- Fine grained (D less than 16 mm)
- Coarse-graded (D greater than or equal to 16 mm)
- Mastic asphalt (*gussasphalt*) and hot rolled asphalt

The mean of the latest 32 analyses should be calculated for each of these groups. By and large, mean deviations exceeding the appropriate values in Table 14.2 indicate a group of nonconforming products (Item 7.4 of the standard applies to them). In these circumstances, suitable corrective measures should be taken; the OCL should be reduced by one level as long as the mean deviation remains outside the permissible range.

14.8.5 Additional Tests of Asphalt Mixture Characteristics

The previous deliberations on FPC use a basic level applied to the standard EN 13108-21 regarding tests of the mixture composition. Meanwhile methods of testing other properties of an asphalt mixture are discussed in Annex D (and are very informative). Collected results should be declared and used to support the procedure of extended validation of type testing.

The SMA mixture characteristics that are tested include the following:

- Content of air voids in compacted samples (% v/v) according to EN 12697-8
- Properties of the binder of RAP (only in case of recycled asphalt in an SMA paving mixture) according to EN 12697-4 and EN 12697-3
  - Penetration at 25°C according to EN 1426
  - Ring and ball (R&B) softening point according to EN 1427

Sampling and testing additional physical properties should be carried out at a frequency selected from the three levels A, B, and C:

Level A—testing every 10,000 tonnes
Level B—testing every 5000 tonnes
Level C—testing every 3000 tonnes

Selection criteria of the levels have not been determined in the standard. Test results should be stored in FPC files.

Some final remarks follow:

- Sampling of a mixture for additional tests should be conducted in accordance with EN 12697-27.

TABLE 14.3
The Comparison of Specification on SMA 10 or SMA 11 according to the rules of EN 13108-5

	Germany	Slovakia	Sweden	Poland	Slovenia	Austria
	TL-Asphalt 2007	Slovakia KLAZ 1/2008	VVTBT Bitumenbundna lager. 2008:113	WT-2 Nawierzchnie Asfaltowe 2008	SIST 13108-5:2008	ONORM B 3584:2006
Designation	SMA 11 S	SMA 11	ABS 11	SMA 11	SMA 11	SMA 11 S1
Thickness of layer, mm	35–40	30–50	24–44	35–50	35–40	30–40
Method of preparing samples	Impact (Marshall)	Impact (Marshall)	Impact (Marshall)	Impact (Marshall)	Impact (Marshall)	Impact (Marshall)
Energy of compaction (strokes/side)	50	50	50	50	50	50
Gradation (% Mass of Passing by Sieve)						
16.0	100	100	100	100	100	100
12.5	—	—	—	—	—	—
11.2	90–100	90–100	90–100	90–100	90–100	90–100
10.0	—	—	—	—	—	—
8.0	50–65	—	35–60	50–65	50–60	48–73
6.3	—	—	—	—	—	—
5.6	35–45	—	24–35	35–45	—	—
4.0	—	30–55	19–30	—	30–40	28–43
2.0	20–30	20–35	—	20–30	20–27	20–30
1.0	—	—	—	—	—	—
0.5	—	11–25	12–24	—	10–16	12–24
0.25	—	—	—	—	—	—

(Continued)

TABLE 14.3 (CONTINUED)

The Comparison of Specification on SMA 10 or SMA 11 according to the Rules of EN 13108-5

	Germany	Slovakia	Sweden	Poland	Slovenia	Austria
0.125	—	—	—	9–17	—	—
0.063	8–12	6–12	9–13	8–12	8–12	6–10
Minimum binder content, Bmin	6.6	6.4	6.0 for 50/70 and 70/100	6.0	6.3	6.2
Minimum void content, Vmin	2.5	2.5	2.0	3.0	2.5	2.0
Maximum void content, Vmax	3.0	4.5	3.5	4.0	4.5	4.0
Minimum void filled with binder, VFBmin	Declared	74	NR	NR	74	NR
Maximum void filled with binder, VFBmax		83	NR	NR	89	NR
Permitted binder drain-off	NR	0.3	NR	0.3	0.6	0.6
Resistance to abrasion by studded tires	NR	NR	NR	NR	NR	NR
Water sensitivity ITSR, %	NR	80	NR	90	NR	NR
Resistance to Permanent Deformation						
Maximum proportional rut depth P	NR	NR	NR	NR	NR	NR
Maximum proportional rut depth, testing in air (PRDair)	Declared	5.0	NR	5.0	5.0	9.0
Maximum wheel tracking slope, testing in air (WTSair)	NR	0.1	NR	0.3	NR	NR

Source: Data from EN 13108-5. Bituminous Mixtures—Material specifications—Part 5: Stone Mastic Asphalt.

Note: NR = no requirement; SMA = stone matrix asphalt.

- Samples should be prepared with the same method as applied for qualification tests of a mixture according to a formula (EN 13108-20); careful attention should be paid to using the same method of compacting samples.

14.9   EXAMPLES OF REQUIREMENTS

For those interested in seeing how the CEN-member states have specified their national requirements for SMA, such a comparison is presented in Table 14.3.

Afterword

A few years ago, at the beginning of the work on this book about SMA, I did not suppose it would take on such imposing proportions. Meanwhile, over the course of work on the publication, it turned out that the quantity of accessible materials on SMA was really spectacular, and the range of SMA-related subjects was enormous. While carrying out a survey on relevant publications, it became noticeable that SMA was still a fascinating asphalt mixture to many process engineers the world over.

All in all, it would be appropriate to finish briefly, in contrast to the content of the book, which might seem to be a bit verbose here and there. Nevertheless, I do hope that it will help its readers comprehend SMA and clear up any problems that might arise.

It is a matter of course that the examples quoted in the book cannot fully correspond with personal experiences of each individual reader. Should anybody like to exchange views about SMA mixture, please get in touch with me by e-mail at sma@road.pl.

References

Airey G.D., Collop A.C., Thom N.H. 2004. Mechanical performance of asphalt mixtures incorporating slag and glass secondary aggregates. *Proceedings of the 8th Conference on Asphalt Pavements for Southern Africa* (CAPSA'04) 12–16 September.

Andersen P.J., Thau M. 2008. 1st generation system for specification and documentation of asphalt surfacings exhibiting noise-reducing properties. *Proceedings of the 4th Euroasphalt & Eurobitume Congress*, Copenhagen.

Anderson D.A. 1987a. Guidelines for use of dust in hot-mix asphalt concrete mixtures. *Journal of the Association of Asphalt Paving Technologists* 56: 492–516.

Anderson D.A. 1987b. Guidelines on the use of baghouse fines. National Asphalt Pavement Association. *Information Series* 101–111.

Anderson D.A., Tarris J.P., Brock J.D. 1982. Dust collector fines and their influence on mixture design. *Journal of the Association of Asphalt Paving Technologists* 51: 363–397.

Arand W. 1996. Asphalt roads under the influence of weather and traffic. *Proceedings of the Eurasphalt & Eurobitume Congress*, Strasbourg.

Asphalt Handbook, The. Manual Series No. 4 (MS-4). Asphalt Institute, Lexington, KY.

Asphalt Review, December 2004: US experts provide advice on SMA. *Asphalt Review* 23(4).

Behbahani H., Nowbakht S., Fazaeli H., Rahmani J. 2009. Effect of fiber type and content on the rutting performance of Stone Matrix Asphalt. *Journal of Applied Sciences* 9(10): 1980–1984.

Behle T., Jannicke B., Radenburg M., Schmidt H. 2005. Mineralstoffkonzepte mit unterschiedlich polierresistenten Splitten und deren Einfluss auf die Griffigkeitsentwicklung von Splittmastixasphalt. Strasse und Autobahn No.8.

Bellin P. 1997. Development, principles, and long-term performance of stone mastic asphalt in Germany. SCI & IAT Joint Seminar, London.

Bennert T., Hanson D., Maher A. 2004. Demonstration Project—The Measurement of Pavement Noise on New Jersey Pavements Using the NCAT Noise Trailer. New Jersey Department of Transportation.

Beuving E., De Jonghe T., Goos D., Lindahl T., Stawiarski A. 2004. Environmental impacts and fuel efficiency of road pavements. Industry report. Eurobitume & EAPA BRUSSELS.

Bolzan P.E. 2002. The rehabilitation of the Ricchieri Highway in Argentina with SMA and thin-SMA technologies. *Proceedings of the 9th International Conference on Asphalt Pavements*, Copenhagen.

Boratyński J., Krzemiński J. 2005. Compaction of SMA mixtures in the laboratory and on building site (Zageszczanie mieszanek mastyksowo-grysowych (SMA) w laboratorium i na drodze - in Polish). Drogownictwo 7–8/2005.

Brennan M.J., Nolan J., Murphy D., Lohan G. 2000. Designing stone mastic asphalt. Road *Materials and Pavement Design* 1(2): 227–243.

Brown D.C. 2002. SMA: Built for the long haul. Highway agencies are showing renewed interest in stone matrix asphalt. *Better Roads Magazine*, October 2002.

Brown E.R., Cooley Jr L.A. 1999. Designing stone matrix asphalt mixtures for rut-resistant pavements. National Cooperative Highway Research Program Report 425. *National Academy Press*, Washington, DC.

Brown E.R., Haddock J.E. 1997. A method to ensure stone-to-stone contact in Stone Matrix Asphalt paving mixtures. *National Center for Asphalt Technology*, Auburn University, Auburn, AL, NCAT Report 97-2.

Brown E.R., Hainin M.R., Cooley A., Hurley G. 2004. Relationship of air voids, lift thickness, and permeability in hot mix asphalt pavements. National Cooperative Highway Research Program. Report 531, Transportation Research Board, Washington, DC.

Brown E.R., Mallick R.B. 1994. Stone matrix asphalt—Properties related to mixture design. *National Center for Asphalt Technology*, Auburn University, Auburn, AL, NCAT Report 94–2.

Bryant P. 2006. Filler: is it 'fixing' your binder? *Asphalt Review* 25(1): March 2006.

Carswell J. 2002. The design and performance of thin surfacing layers. *Proceedings of the 9th International Conference on Asphalt Pavements*, Copenhagen.

Carswell J. 2004. The design and performance of thin surfacing layers. *Asphalt Review* 23(1).

Celaya B.J., Haddock J.E. 2006. Investigation of coarse aggregate strength for use in stone matrix asphalt. Civil Engineering Joint Transportation Research Program, School of Civil Engineering Purdue University, Report FHWA/IN/JTRP-2006/4. West Lafayette, IN

Chen J.S, Pen C-H. 1998. Analyses of tensile failure properties of asphalt-mineral filler mastics. *Journal of Materials in Civil Engineering* 10(4): 256–262.

Chen J.S., Lin K-Y., Young S-Y. 2004. Effects of crack width and permeability on moisture-induced damage of pavements. *Journal of Materials in Civil Engineering* 16(3): 276–282.

Choubane, B., Page G.C., Musselman J.A. 1998. Investigation of water permeability of coarse graded superpave pavements. *Journal of the Association of Asphalt Paving Technologists* 67: 254–276.

Compaction and laying of asphalt pavements. Theory and practice. Dynapac 2004.

Cooley L.A. Jr., Brown E.R. 2003. Potential of using stone matrix asphalt (SMA) for thin overlays. *National Center for Asphalt Technology*, Auburn University, Auburn, AL, NCAT Report 03-01. April.

Cooley L.A. Jr., Prowell B.D., Brown E.R. 2002. Issues pertaining to the permeability characteristics of coarse-graded superpave mixes. Association of Asphalt Paving Technologists 2002, *National Center for Asphalt Technology*, Auburn University, Auburn, AL, NCAT Report 2002-06.

Daines M.E. 1985. Cooling of bituminous layers and time available for their compaction. Research Report No. 4, 1985. Transport and Road Research Laboratory, U.K., Crowthorne, Berkshire.

Damm K.W. 2002. Asphalt flow improvers as intelligent filler for hot asphalt—A new chapter in asphalt technology. *Journal of Applied Asphalt Binder Technology* 2(1): 36–69.

Damm K.W. Harders O. 2000. Long-term performance of bridge deck surfacing for heavy traffic. *Proceedings of the 2nd Eurasphalt & Eurobitume Congress*, Barcelona.

di Benedetto H., De La Roche C., Francken L. 1997. Fatigue of bituminous mixes: Different approaches and RILEM interlaboratory tests. *Proceedings of the 5th International RILEM Symposium Mechanical Test for Bituminous Materials*. Lyon, France.

Drüschner L., Els H., Erhardt H. et al. 2001. Asphaltdeckschichten mit anforderungsgerechter Griffigkeit. Maßnahmen zur Planung und Ausführung. Deutscher Asphaltverband e.V. (DAV).

Drüschner L. 2005. The German origin of SMA. *Asphalt Review* 24(3): November 2005.

Drüschner L. 2006. The German origin of SMA. *Asphalt Review* 25(1): March 2006.

Drüschner L., Schäfer V. 2000. Splittmastixasphalt. DAV Leitfaden. Deutscher Asphaltverband.

EAPA. Airfield uses of asphalt. *European Asphalt Pavement Association*. Ref: (3)2-03-00.015. May 2003, Breukelen (The Netherlands).

EAPA. Heavy duty surfaces. The arguments for SMA. European Asphalt Pavements Association. 1998 (ISBN 90-801214-8-7), Breukelen (The Netherlands).

Fabb T.R.J. 1973. The influence of mix composition, binder properties and cooling rate on asphalt cracking at low temperatures. *Journal of the Association of Asphalt Paving Technologists* 43: 285–331.

Factors affecting HMA permeability. Washington State Department of Transportation. Materials Laboratory. Tech Notes February 2005.

Ferguson A., Fordyce D., Khweir K. 1999. Designing Stone Structure Wearing Course Mixtures. *Proceeding of the Third European Symposium on Performance and durability of bituminous Material and Hydraulic Stabilised Composites, Leeds April 1999*. Editors J G Cabrera and S E Zoorob, AEDIFCATIO publishers, D-79104 Freiburg i. Br. And CH-8103 Unterengstringen/ Zurich.

Fowler D.W., Allen J.J., Lange A., Range P. 2006. The prediction of coarse aggregate performance by micro-deval and other aggregate tests. International Center for Aggregates Research, The University of Texas at Austin. Austin, Tx, Research Report ICAR 507-1F.

Francken L., Vanelstraete A. 1993. New developments in analytical asphalt mix design. *Proceeding of Eurobitume Congress* Stockholm, Sweden.

Frankfurt takes off from asphalt. Low temperature asphalt makes reconstruction of the airport possible. Paper of the Sasolwax GmbH, undated.

Gardziejczyk W. 2002. The texture of road pavements—test methods, assesment indicators and its influence on rolling noise (Tekstura nawierzchni drogowych—metody pomiaru, wskaźniki oceny i jej wpływ na hałas toczenia - in Polish). Drogi i Mosty, Vol.2.

Gardziejczyk W., Wasilewska M. 2003. The influence of aggregate on antiskid properties of road pavements (Wpływ kruszywa na właściwości przeciwpoślizgowe nawierzchni drogowych - in Polish). Drogownictwo 11, 2003, pp.347–353.

Gärtner K., Graf K., Meyer D., Scheuer S. 2006. Lärmtechnisch optimierte Splittmastixasphaltdeckschichten. Strasse und Autobahn, 12/2006.

Gärtner K., Graf K., Schünemann M. 2009. Asphaltbinderschichten nach den Splittmastixprinzip. Strasse und Autobahn, 7/2009.

Grabowski W., Wilamowicz J. 2001. Testing of mineral fillers' structure influence on their functional properties (in Polish). *Proceedings of the 7th International Conference Durable and Safe Road Pavements*. Kielce (Poland).

Graf K. 2006. Splittmastixasphalt—Anwendung und Bewährung. Rettenmaier Seminar eSeMA'06. Zakopane (Poland).

Gudimettla J.M., Cooley L.A. Jr., Brown E.R. 2003. Workability of hot mix asphalt. *National Center for Asphalt Technology*, Auburn University, Auburn, AL, NCAT Report 03-03.

Harris B.M., Stuart K.D. 1995. Analysis of mineral fillers and mastics used in stone matrix asphalt. *Journal of the Association of Asphalt Paving Technologists* 64: 54–95.

Hensley J. Eliminate twelve segregation snarls. *Asphalt Institute* web site, accessed 20 December 2009. http://www.asphaltinstitute.org/public/engineering/PDFs/Pavement_ Performance/Eliminate_12_segregation_snarls.pdf

Hicks R.G., Leahy R.B., Cook M., Moulthrop J.S., Button J. 2003. Road map for mitigating national moisture sensitivity concerns in hot-mix pavements. Moisture sensitivity of asphalt pavements. TRB National Seminar, San Diego, California, on 4–6 February 2003.

Höbeda P. 2000. Testing the durability of asphalt mixes for severe winter conditions. *Proceedings of the 2nd Eurasphalt & Eurobitume Congress*, Barcelona.

Hot-mix asphalt paving Handbook 2000. 2nd Edition, *US Army Corps of Engineers*, Washington, DC. (ISBN 0309071577).

Hunter R.N. 1994. Bituminous mixtures in road construction. Thomas Telford 1994. London ISBN 0727716832.

Huschek S. 2004. Polierwiderstand und Griffigkeit - Langzeiterfahrungen. Strasse und Autobahn 9/2004.

Hutschenreuther J., Woerner T. 1998. Asphalt im Strassenbau. Verlag fuer Bauwesen. 1998 Berlin, ISBN 3345006138.

Huurman M., Medani T.O., Scarpas A., Kasbergen C. 2003. Development of a 3D-FEM for surfacings on steel deck bridges. *International Conference on Computational & Experimental Engineering and Sciences*, ICCES'03 Corfu, Greece.

Huurman M. 2000. Cyclic triaxial tests on asphalt concrete related to rutting. *Proceedings of the 2nd Eurasphalt & Eurobitume Congress*, Barcelona.

Isacsson U., Vinson T.S., Zeng H. 1997. The influence of material factors on the low temperature cracking of asphalt mixtures. *Proceedings of the 5th International RILEM Symposium Mechanical Test for Bituminous Materials*, Lyon, France.

Iwański M. 2003. The influence of hydrated lime on asphalt concrete properties. *Proceedings of the 9th International Conference Durable and Safe Road Pavements*. Kielce (Poland) (in Polish)

Jabłoński K. 2000. Stone-mastic mixture as a wearing course on highways. The information on standard ZN-71/MK-CZDP-3. (Mieszanka mastyksowo-grysowa jako warstwa ścieralna na autostradach. Informacja o normie ZN-71/MK-CZDP-3—in Polish). Polskie Stowarzyszenie Wykonawców Nawierzchni Asfaltowych, Seminarium Wybrane zagadnienia technologiczno-materiałowe budowy asfaltowych nawierzchni płatnych autostrad. Warszawa (Poland).

Jacobs M., Hopman P., Molenaar A. 1996. The crack growth mechanism in asphaltic mixes. *Proceedings of the Eurasphalt & Eurobitume Congress*, Strasbourg.

Jacobs M.M.J., Fafie J.J. 2004. Improvement of the early life skid resistance of stone mastic asphalt by gritting or sanding. *Proceedings of the 3rd Eurasphalt & Eurobitume Congress*, Vienna.

Johnson E.N., Marasteanu M.O., Clyne T.R., Li X. 2004. Validation of superpave fine aggregate angularity values. Minnesota Department of Transportation. University of Minnesota Department of Civil Engineering, Minneapolis, MN, Report No. MN/RC—2004-30.

Joubert P.B., Gounder L., van Wyk S. 2004. Experimental asphalt sections in the runway touch down zone on Johannesburg International Airport. *Proceedings of the 8th Conference on Asphalt Pavements for Southern Africa (CAPSA'04)*.

Judycki J. 1990. Bending test of asphaltic mixtures under static loading. *Proceedings of the 4th International Conference RILEM in Budapest*, Chapman and Hall, London.

Judycki J., Jaskuła P. 1999. Testing of asphalt concrete with hydrated lime on water and freeze-thaw resistance (Badania betonu asfaltowego z wapnem hydratyzowanym—in Polish). *Proceedings of the 5th International Conference Durable and Safe Road Pavements*. Kielce (Poland).

Judycki J., Pszczoła M. 2002. The influence of binder type and content and type of asphalt mixture on low temperature cracking (Wpływ rodzaju i zawartości asfaltu oraz typu mieszanki mineralno-asfaltowej na spękania niskotemperaturowe—in Polish). *Proceedings of the 8th International Conference Durable and Safe Road Pavements*. Kielce (Poland).

Kandhal P.S., Cooley L.A. Jr. 2003. Accelerated laboratory rutting tests: Evaluation of the Asphalt Pavement Analyzer. National Cooperative Highway Research Program. Washington, DC, Report 508.

Kandhal P.S., Lynn C.Y., Parker, Jr. F. 1998. Characterization tests for mineral fillers related to performance of asphalt paving mixtures. National Center for Asphalt Technology, Auburn University, Auburn, AL, NCAT Report No. 98-2.

Kandhal P.S., Mallick R.B., Brown E.R. 1998. Hot mix asphalt for intersections in hot climates. National Center for Asphalt Technology, Auburn University, Auburn, AL, NCAT Report 98-06.

Kandhal, P.S. 1981. Evaluation of baghouse fines in bituminous paving mixtures. *Journal of the Association of Asphalt Paving Technologists* 50: 150–210.

Kanitpong K., Bahia H.U. 2003. Evaluation of the extent of HMA moisture damage in Wisconsin as it relates to pavement performance. Wisconsin Department of Transportation. Madison, WI.

Kelkka M., Valtonen J. 2000. Test road for permanent deformation. *Proceedings of the 2nd Eurasphalt & Eurobitume Congress*, Barcelona.

Krans R.L., Tolman F., Van de Ven M.F.C. 1996. Semi-circular bending test: a practical crack growth test using asphalt concrete cores. In: Reflective Cracking in Pavements: Design and Performance of Overlay Systems (Edited by Francken L., Beuving E., and Molenaar A.A.A.). *Proceedings of the 3rd International RILEM Conference. E & FN SPON*, 1996 London, ISBN 041922260X.

Kreide M. 2000. Evaluation of Normal Paving Grades and Polymer Modified Binders in Germany (ARBIT—Quality Criteria). *Proceedings of the 6th International Conference, Durable and Safe Road Pavements*, Kielce (Poland).

Kucharski R. 1979. Traffic noise (Hałas drogowy—in Polish), Wydawnictwo Naukowe PWN, Warszawa.

Larsen O.R. 2002. SMA at the new Oslo Airport Gardermoen. Presentation at the 1st International Workshop SMA and Fibres VIATOP, Hannover, 12th November.

Lees G. 1969. The rational design of aggregate grading for dense asphaltic compositions. *Journal of the Association of Asphalt Paving Technologists* 39: 60–97.

Little D.N., Epps J. 2001. The benefits of hydrated lime in hot mix asphalt. Report for National Lime Association.

Marasteanu M.O., Li X., Clyne T.R., Voller V.R., Timm D.H., Newcomb D.E. 2004. Low temperature cracking of asphalt concrete pavements. University of Minnesota, Minneapolis, Report MN/RC—2004-23 Minnesota Department of Transportation, St. Paul.

Martin J.S., Cooley A.Jr., Hainin H.R. 2003. Production and construction issues for moisture sensitivity of hot-mix asphalt pavements. Moisture sensitivity of hot-mix asphalt pavements. TRB National Seminar, San Diego, California.

Medani T.O. 2001a. Asphalt surfacing applied to orthotropic steel bridge decks. A literature review. Report 7-01-127-1. Delft University of Technology, Delft.

Medani T.O. 2001b. Towards a new design philosophy for asphalt surfacings on orthotropic steel decks. Report 7-01-127-2. Delft University of Technology, Delft.

Milster R., Emperhoff W., Graf K., Lips C., Mansfeld R. 2004. Ratschläge für den Einbau von Walzasphalt. Deutscher Asphaltverband e.V. (DAV) Bonn.

Molenaar, J.M.M. and Molenaar, A.A.A. 2000. Fracture toughness of asphalt in the semi-circular bend test. *Proceedings of the 2nd Eurasphalt & Eurobitume Congress*, Barcelona.

Muniandy R., Aburkaba E.E., Hamid H.B., Yunus R.BT. 2009. An initial investigation of the use of local industrial wastes and by-products as mineral fillers in stone mastic asphalt pavements. *ARPN Journal of Engineering and Applied Sciences*, 4(3).

Nicholls J.C., Carswell I., James D. 2008. Performance data on thin surfacing systems after up to 15 years in service. *Proceedings of the 4th Eurasphalt and Eurobitume Conference*, Copenhagen.

Nolle B. Durable skid resistant asphalt roads. Technical principles and contractual aspects. Asphalt 5/2004.

Obert S. 2000. Predicting the performance of stone mastic asphalt. Young Researchers Forum, London.

Oliver J. 2001. On the skids—The friction factor. *Asphalt Review* 20(3).

Olszacki J. 2005. The determination of the water permeability and noise absorption of asphalt concrete used in porous courses (Określenie wodoprzepuszczalności i dz´więkochłonności betonoasfaltów stosowanych w nawierzchniach drenujących), Ph.D. thesis, Kielce University of Technology (Poland).

Ordens R.A.P., Cenek P., Descornet G., Fwa T.F., Gothié M., Schmidt B., Wambold J. 1999. Friction and texture measurement methods. *Proceedings of the XXI World Road Congress*, World Road Association, Kuala Lumpur.

Pandyra W., Błażejowski K., Styk S. 1994. The end of surface dressings? (Koniec powierzchniowych utrwaleń?—in Polish). Drogownictwo 9/1994.

Pashula C. 2005. Implementation of SMA in Australia. *Asphalt Review* 24(3).

Perez F., Rodriguez M., De Visscher J., Vanelstraete A., De Bock L. 2004. Design and performance of hot mix asphalts with high percentages of reclaimed asphalt: Approach followed in the Paramix project. *Proceedings of the 3rd Eurasphalt & Eurobitume Congress*, Vienna.

Pierce L.M., Willoughby K.A., Uhlmeyer J.S., Mahoney J.P., Anderson K.W. 2002. Temperature and density differentials in asphalt concrete pavement. *Proceedings of the 9th International Conference on Asphalt Pavements*, Copenhagen.

Pretorius F.J., Wise J.C., Henderson M. 2004. Development of application differentiated ultra-thin asphalt friction courses for southern African application. *Proceedings of the 8th Conference on Asphalt Pavements for Southern Africa (CAPSA'04)*.

Prowell B.D., Watson D.E., Hurley G.C., Brown E.R. 2009. Evaluation of stone matrix asphalt (SMA) for airfield pavements. Auburn University, Auburn, AL, *Airport Asphalt Pavement Technology Program Association of Asphalt Paving Technologists*, P 04-04, February.

Radenberg M. 1997. Modifizierte Walzasphalte im Spurbildungstestvergleich. Asphalt, Heft 4.

Read J., Whiteoak D. 2003. The Shell bitumen handbook. Thomas Telford 2003, London ISBN 072773220.

Renken P. 2000. Influence of specimen preparation onto the mechanical behaviour of asphalt aggregate mixtures. *Proceedings of the 2nd Eurasphalt & Eurobitume Congress*, Barcelona.

Renken P. 2004. The compaction resistance of asphalt mixes—A comprehensive performance related property. *Proceedings of the 3rd Eurasphalt & Eurobitume Congress*, Vienna.

Report FHWA-IF-03-019. Fly ash facts for highway engineers. American Coal Ash Association, Aurora, CO.

Richardson J.T.G. 1997. Stone mastic asphalt in the UK. Symposium on Stone Mastic Asphalt and Thin Surfacings, London.

Richter E. 1997. Kompaktasphalt. *Proceedings of the 3rd International Conference Durable and Safe Road Pavements*. Kielce (Poland).

Rigo J.M. 1993. General introduction, main conclusions of the 1989 conference on reflective crackings in pavements and future prospects. *Proceedings of the 2nd International RILEM Conference*, Liege, Belgium.

Roberts, F.L, Kandhal, P.S., Brown, E.R, Lee D.Y., Kennedy T.W. 1996. Hot mix asphalt materials, mixture design, and construction. NAPA Research and Education Foundation, Lanham, MD.

Roovers M.S., de Graaf D.F., van Moppers R.K.F. 2005. Round-robin test rolling resistance/ energy consumption. Report of The Road and Hydraulic Engineering Division of Rijkswaterstaat, Delft, No. DWW-2005-046.

Said S., Jacobson T., Hornwall F., Wahlström J. 2000. Evaluation of permanent deformation in bituminous mixes. *Proceedings of the 2nd Eurasphalt & Eurobitume Congress*, Barcelona.

Said S., Wahlström J. 2000. Validation of indirect tensile method for fatigue characterizing of bituminous mixes. *Proceedings of the 2nd Eurasphalt & Eurobitume Congress*, Barcelona.

Sandberg T. 2001. Heavy Truck modelling for fuel consumption. Simulations and measurements. Linköping University, Linköping, Sweden.

Sandberg U. 2001. Noise emissions of road vehicles effect of regulations final report 01-1. I-ince working party on noise emissions of road vehicles (WP-NERV). International Institute of Noise Control Engineering. Linköping, July 2001, p. 42.

Sandberg U., Ejsmont J.A. Tyre/road noise reference handbook. Informex, Sweden, 1999 Kisa ISBN 9163126109.

Santucci L. 2002. Moisture sensitivity of asphalt pavements. Institute of Transportation Studies, UC Berkeley, Technical Topics.

Schabow J. 2005. Das neue Merkblatt für das Verdichten von Asphalt. Strasse und Autobahn, 8/2005, pp.453–458.

Schellenberg K., von der Weppen W. 1986. Verfahren zur Bestimmung der Homogenitäts-Stabilität von Splittmastixasphalt. Bitumen 1/1986, pp.13–14.

Schellenberger W. 1997. Einbau von Asphaltbeton und der Einfluss der Witterungsbedingungen. Asphalt 7–8/1997, pp.29–32.

Schellenberger W. 2002. Beurteilung der Eignung unterschiedlicher Füller für den Asphaltstrassenbau. Bitumen 1/2002, pp.6–13.

Scherocman, J.A. 1991. Stone mastic asphalt reduces rutting. *Better Roads* 61(11): 1991.

Schmiedlin R.B., Bischoff D.L. 2002. Stone matrix asphalt. *The Wisconsin experience. Wisconsin Department of Transportation.* Madison, WI.

Schroeder I., Kluge H.J. 1992. Erfahrungen mit Splittmastixasphalt. Bitumen 4/1992.

Schroer J. 2006. Technician Issues Missouri. NCAUPG Workshop (presentation). St. Louis, MO.

Schünemann M. 2006. Splittmastixbinder Eine bessere Alternative zum «Asphaltbetonbinder»? Rettenmaier Seminar eSeMA'07, Zakopane (Poland).

Schünemann M. 2007. Faserqualität. Eine wesentliche Voraussetzung zum Herstellen von qualitätsgerechten Asphaltbefestigungen. Rettenmaier Seminar eSeMA'07, Zakopane (Poland).

SHRP-A-404. 1994. Fatigue response of asphalt-aggregate mixes. Asphalt research program. Institute of Transportation Studies, University of California, Berkeley. Strategic Highway Research Program. National Research Council. Washington, DC.

Śliwiński J. 1999. Cement concrete—Design and properties (Beton zwykły—projektowanie i podstawowe właściwości—in Polish). Krakow, Polski Cement.

Sluer B.W. 2001. Kjellbase een toekomst zonder sporen? (in Dutch) Asfalt nr. 2.

Sluer B.W. 2002. Kjellbase. A future without ruts. Development of a heavy duty pavement. Presentation at the 1st International Workshop. SMA and JRS Fibers. Hannover.

Sluer B.W., Waanders G.W.J., Smit H.J.J., Gouw S., undated. Steinskelettgemische in Tragschichten. Ein Sieg im Kampf gegen die Bildung von Spurillen? (in German) Undated.

Spuziak W. 2000, Conditions of transport of hot bituminous mixtures (Warunki transportu gorących mas bitumicznych—in Polish). *Proceedings of the 6th International Conference Durable and Safe Road Pavements.* Kielce (Poland).

Stakston A.D., Bahia H. 2003. The effect of fine aggregate angularity, asphalt content, and performance graded asphalts on hot mix asphalt performance. WisDOT Highway Research Study 0092-45-98. Wisconsin DoT, Madison, WI.

Steernberg K., Read J.M., Seive A. 2000. Fuel resistance of asphalt pavements. *Proceedings of the 2nd Eurasphalt & Eurobitume Congress,* Barcelona.

Stephenson G., Bullen F. 2002. The design, creep, and fatigue performance of stone mastic asphalt. *Proceedings of the 9th International Conference on Asphalt Pavements,* Copenhagen.

Stroup-Gardiner M., Law M., Nesmith C. 2000. Using infrared thermography to detect and measure segregation in hot mix asphalt pavements. *International Journal of Pavement Engineering* 1(4):265–284.

Superpave Mixture Design Guide. WesTrack Forensic Team Consensus Report. 2001 U.S. Department of Transportation, FHWA,Washington, DC.

Sybilski D., Horodecka R. 1998. Assessment of rutting resistance of asphalt mixtures and relationship to the test method and used binder type. (Ocena odporności na koleinowanie mieszanek mineralno-asfaltowych w zależności od zastosowanego lepiszcza i metody badania—in Polish). *Proceedings of the 4th International Conference Durable and Safe Road Pavements,* Kielce (Poland).

Sybilski D., Mechowski T. 1996. Evaluation of durability of bituminous mixtures. *Proceedings of the 1st Eurasphalt & Eurobitume Congress*, Strasbourg.

Sybilski D., Styk S. 1996. Choice of binder for hot thin layers. *Proceedings of the 1st Eurasphalt and Eurobitume Congress*, Strasbourg.

Tangella R.S.C.S., Craus J., Deacon J.A., Monismith C.L. 1990. Summary report on fatigue response of asphalt mixtures. Strategic Highway Research Program, Project A-003-A. Institute of Transportation Studies, University of California, Berkeley, February.

Tayebali A.A., Khosla N.P., Malpass G.A. 1996. Impact of fines on asphalt mix design. The Center for Transportation and the Environment, North Carolina State University, Chapel Hill, NC.

Tuckett G.M., Jones G.M., Littlefield G. 1969. The effect of mixture variables on thermally induced stresses in asphaltic concrete. *Journal of the Association of Asphalt Paving Technologists* 39: 703–744.

Ulmgren N. 1996. Functional testing of asphalt mixes for permanent deformation by dynamic creep test. Modification of method and round-robin test. *Proceedings of the 1st Eurasphalt & Eurobitume Congress*, Strasbourg.

Ulmgren N. 2000. Temperature scanner—An instrument to achieve a homogenous asphalt pavement. *Proceedings of the 2nd Eurasphalt & Eurobitume Congress*, Barcelona.

Ulmgren N. 2004. The influence of the mastic on the durability of asphalt pavements as studied by the shaking abrasion test. *Proceedings of the 3rd Eurasphalt & Eurobitume Congress*, Vienna.

Utterodt R., Egervari R. 2009. Effectiveness of the Compactasphalt® technology. Thirteenth International Flexible Pavements Conference "PAVEMENTS FOR TODAY", Gold Coast, Australia.

Valtonen J., Hyyppa I., Sainio P. 2002. Noise reduction vs wearing properties. *Proceedings of the 9th International Conference on Asphalt Pavements*, Copenhagen.

van de Ven M.F.C., Voskuilen J.L.M., Tolman F. 2003. The spatial approach of hot mix asphalt. *Proceedings of the 6th RILEM Symposium PTEBM'03*. Zurich.

Viman L., Wendel M., Said S.F. 2004. Effect of flow mixing technique, KGO-III, on characteristics of bituminous surfacing. *Proceeding of the 3rd Eurasphalt & Eurobitume Congress*, Vienna.

von Brochove G., Hamzah M.O. 2008. The state of the art in the field of silent road surfaces. *Proceedings of the 4th Euroasphalt & Eurobitume Congress*, Copenhagen 2008, paper 402–101.

von Brochove G.G., Voskuilen J., Visser A.F.H.M. 2008. A new type asphalt surface layer for steel bridge decks. *Proceedings of the 4th Eurasphalt & Eurobitume Conference*, Copenhagen, paper 402–102.

Voskuilen J.L.M. 2000. Ideas for a volumetric mix design method for stone mastic asphalt. *Proceedings of the 6th International Conference Durable and Safe Road Pavements*, Kielce (Poland).

Voskuilen J.L.M., Jacobs M.M.J., van Bochove G.G. 2004. Determination of the quality of the coarse crushed aggregates for SMA during mix design, type testing and factory production control. *Proceedings of the 3rd Eurasphalt & Eurobitume Congress*, Vienna, paper 326, p. 1802.

Wegan V. 2000. Surfacing of concrete bridges. Danish Road Directorate, Danish Road Institute, Roskilde, Denmark, Report 106. December.

White T.D., Haddock J.E., Rismantojo E. 2006. Aggregate tests for hot-mix asphalt mixtures used in pavements. National Cooperative Highway Research Program, NCHRP Report 557, Transportation Research Board, Washington, DC.

Willoughby K.A., Mahoney J.P, Pierce L.M., Uhlmeyer J.S., Anderson K.W., Read S.A., Muench S.T., Thompson T.R., Moore R. 2001. Construction-related asphalt concrete pavement temperature differentials and the corresponding density differentials. Washington State Transportation Center (TRAC). Seattle, WA.

WsDOT. 2005. Factors affecting HMA permeability. Washington State Department of Transportation. Materials Laboratory. Tech Notes.

Zichner G. 1971. Wearing courses of stone and mastic on pavements. US Patent No. 3797951.

Zichner G. 1972, MASTIMAC unad MASTIPHALT bituminöse Gemische für hochwertige Deckschichten. STRABAG Schriftenreihe 8, Folge 4.

Guidelines & Regulations

150/5370-10B. U.S. Department of Transportation. Federal Aviation Administration. Advisory Circular AC No: 150/5370-10B. Standards for specifying construction of airports, 2005, (Table 4, pp. P-401–411)

ETL 04-8. Stone matrix asphalt (SMA) for Air Force pavements. Engineering Technical Letter (ETL) 04-8, Department of the Air Force, Headquarters Air Force Civil Engineer Support Agency. 2004

FGSV Arbeitspapier Nr. 42. Prüfung und Kennzeichnung von stabilisierenden Zusätzen und stabilisierenden Stoffen für den Asphaltstraßenbau. Ausgabe 1997 (in German)

HA MCHW. Highways Agency. Manual of contract documents for highway works. Volume 1. Specification for highway works. Series 900. Road pavements—bituminous bound materials. Amendment—November 2008. United Kingdom

KLAZ 1/2008. Katalógové Listy Asfaltových Zmesí (doplnok k platným TKP). Ministerstvo dopravy, pôšt a telekomunikácií Slovenskej Republiki. Sekcia cestnej dopravy a pozemných komunikácií. Apríl 2008 (in Slovak)

KLK 1/2009. Katalógové Listy Kameniva (doplnok k platným TKP). Ministerstvo dopravy, pôšt a telekomunikácií Slovenskej Republiki. Sekcia cestnej dopravy a pozemných komunikácií. Apríl 2009 (in Slovak)

M VA 2005. Merkblatt für das Verdichten von Asphalt (M VA), Ausgabe 2005, FGSV Köln, Germany. 2005

NAPA QIS 122. Designing and constructing SMA mixtures—State of the practice. National Asphalt Pavements Association, Lanham, MD, Quality Improvement Series 122. 2002

NAS AAPA 2004. National asphalt specification. 2nd Edition. Australian Asphalt Pavement Association (AAPA). April 2004

OPSS.MUNI 1151. Material specification for superpave and stone mastic asphalt mixtures. Ontario Provincial Standard Specification and Appendix 1151-A, Commentary for OPSS.MUNI 1151, Ontario Provincial Standards for Roads and Public Works (OPS) November 2006

PANK 2000. Finnish asphalt specification 2000. Finnish Pavement Technology Advisory Council (PANK). Helsinki 2000

PD 6682-2:2003. Aggregates—Part 2: Aggregates for bituminous mixtures and surface treatments for roads, airfields and other trafficked areas—Guidance on the use of BS EN 13043 British Standards Institution, London 2003.

PD 6691:2007 Guidance on the Use of BS EN 13108 Bituminous mixtures—Material specifications. British Standards Institution, London 2007

RVS 08.16.01. Anforderungen An Asphaltschichten. Technische Vertragsbedingunge Bituminöse Trag- und Deckschichten. Ausgabe 1. Jänner 2007 (in German) Österreichische Forschungsgesellschaft Strasse–Schiene–Verkehr, Vien, Austria

RVS 08.97.05:2007. Technische Vertragsbedingungen. Baustoffe. Anforderungen An Asphaltmischgut RVS 08.97 .05, Ausgabe 1 . Jänner 2007 (in German)

Technicke Kvalitativni Podminky Staveb Pozemnich Komunikaci. Kapitola 7. Hutnene Asfaltove Vrstvy. Praha 1999 (in Czech)

TL Asphalt-StB 07. Technische Lieferbedingungen für Asphaltmischgut für den Bau von Verkehrsflächenbefestigungen. Ausgabe 2007 (in German) Forschungsgesellschaft für Straßen- und Verkehrswesen, FGSV, Köln

TL Gestein StB 04. Technische Lieferbedingungen für Gesteinskörnungen im Straßenbau, Ausgabe 2004 (in German) Forschungsgesellschaft für Straßen- und Verkehrswesen, FGSV, Köln

TL PmB 01. Technische Lieferbedingungen für gebrauchsfertige polymermodifizierte Bitumen. Ausgabe: 2001 (in German) Forschungsgesellschaft für Straßen- und Verkehrswesen, FGSV, Köln

TP 109, zm.c.1. 2000. Asfaltove hutnene vrstvy se zvysenou odolnosti proti tvorbe trvalych deformaci (in Czech)

TP A-08. Technical test specifications for asphalt in road building TP A-StB. Part 8: Volumetric characteristics of bituminous specimens and compaction (TP A-08). Forschungsgesellschaft für Straßen- und Verkehrswesen, FGSV, Köln 2007

UFGS-32 13 17 (August 2008). Unified facilities guide specifications. Stone matrix asphalt (SMA) for airfield pavements. USACE/NAVFAC/AFCESA/NASA. 2008

U.S. Department of Transportation. Federal Aviation Administration. Advisory Circular AC No: 150/5320-6E. Airport pavement design and evaluation. Washington, DC, September 2009

ÚT 2-3.301-5:2008 ÚTŰGYI MŰSZAKI ELŐÍRÁS. Útépítési aszfaltkeverékek. Zúzalékvázas masztixaszfalt (SMA). Gazdasági És Közlekedési Minisztérium Hálózati Infrastruktúra Föosztálya. 2008 (in Hungarian)

VVTBT Bitumenbundna lager.VV Publ 2008:113. Vägverket 2008 (in Swedish)

WT-1. Wytyczne Techniczne WT-1 Kruszywa 2008. Instytut Badawczy Dróg i Mostów 2008 (in Polish)

WT-2. Wytyczne Techniczne WT-2. Nawierzchnie Asfaltowe 2008. Instytut Badawczy Dróg i Mostów 2009 (in Polish)

ZTV Asphalt - StB 01. Zusätliche Technische Vertragsbedingungen und Richtlinien fuer den Bau von Fahrbahndecken aus Asphalt. Ausgabe 2001 (in German) Forschungsgesellschaft für Straßen- und Verkehrswesen, FGSV, Köln

ZTV Asphalt - StB 07. Zusätliche Technische Vertragsbedingungen und Richtlinien fuer den Bau von Verkehrsflächenbefestigungen aus Asphalt. Ausgabe 2007 (in German) Forschungsgesellschaft für Straßen- und Verkehrswesen, FGSV, Köln

ZTV bit-StB 84. Zusäzliche Technische Vertragsbedingungen und Richtlinien für den Bau von Fahrbahndecken aus Asphalt (Additional technical contract terms and guidelines for the construction of asphalt surfacings) (in German) Forschungsgesellschaft für Straßen- und Verkehrswesen, FGSV, Köln

ZW-SMA-2001. Zasady wykonywania nawierzchni z mieszanki SMA. Instrukcja IBDiM, Zeszyt nr 62. 2001 (in Polish)

Internet

HAUC 2009. Highway Authorities and Utilities Committee, Notes for Guidance on the Use of Stone Mastic Asphalt. Stone Mastic Asphalt—Additional Advice 22nd April 2004. http://www.hauc-uk.org.uk/category/3/pageid/53/newsid/8/ (accessed December 20 2009)

Häusler F.M., Arand W. Süd—Autobahn A2. Deckenausbau mit einem Splittmastixasphalt und NAF 501 Siloware. Asphaltstrassenbau. http://www.trinidad-lake-asphalt.de (accessed 6 December 2009)

RETTENMAIER 2009a. The first mastic treatment. http://www.sma-viatop.com/SMAviatop_ engl/sma_entwicklung/ mastixbehand.shtml?navid=17 (accessed November 10 2009)

RETTENMAIER 2009b. The first Mastimac wearing course. http://www.sma-viatop.com/ SMAviatop_engl/sma_entwicklung/mastimac.shtml?navid=18 (accessed November 10 2009)
SEHAUC 2009. Notes for Guidance on the Use of Stone Mastic Asphalt (SMA) in Hand Lay Situations. The South East Highway Authorities and Utilities Committee. http://www. sehauc.org.uk/notes_SMA.html (accessed December 20 2009)

Standards

AASHTO M325 Standard specification for designing Stone Matrix Asphalt, (former MP 8-00)
AASHTO MP8 Standard Specification for Designing Stone Matrix Asphalt (SMA)
AASHTO R46 Standard practice for designing Stone Matrix Asphalt (SMA), (former PP 41)
AASHTO T 19 Standard method of test for bulk density ("unit weight") and voids in aggregate
AASHTO T 245 Standard method of test for resistance to plastic flow of bituminous mixtures using Marshall apparatus
AASHTO T 283 Resistance of compacted bituminous mixture to moisture-induced damage
AASHTO T 304 Uncompacted void content of fine aggregate
AASHTO T 305 Determination of draindown characteristics in uncompacted asphalt mixtures
AASHTO T 305-97 Standard method of test for determination of draindown characteristics in uncompacted asphalt mixtures
AASHTO T 312 Standard method of test for preparing and determining the density of hot mix asphalt (HMA) specimens by means of the Superpave gyratory compactor
ASTM D 6926-04 Standard practice for preparation of bituminous specimens using Marshall apparatus
ASTM D3387-83(2003) Standard test method for compaction and shear properties of bituminous mixtures by means of the U.S. Corps of Engineers gyratory testing machine (GTM)
ASTM D4013-09 Standard practice for preparation of test specimens of bituminous mixtures by means of gyratory shear compactor
BS 594987:2007 British Standard. Asphalt for roads and other paved areas—specification for transport, laying, and compaction and type testing protocols. Annex I
ČSN 73 6121 (736121) Stavba vozovek - Hutněné asfaltové vrstvy—Provádání a kontrola shody. 1994
EN 1097-4 Tests for mechanical and physical properties of aggregates—Part 4: Determination of the voids of dry compacted filler
EN 12591 Bitumen and bituminous binders—Specifications for paving grade bitumens
EN 12697-1 Bituminous mixtures—Test methods for hot mix asphalt—Part 1: Soluble binder content
EN 12697-2 Bituminous mixtures—Test method for hot mix asphalt—Part 2: Determination of particle size distribution
EN 12697-5 Bituminous mixtures—Test methods for hot mix asphalt—Part 5: Determination of the maximum density
EN 12697-6 Bituminous mixtures—Test methods for hot mix asphalt—Part 6: Determination of bulk density of bituminous specimens
EN 12697-7 Bituminous mixtures—Test methods for hot mix asphalt—Part 7: Determination of bulk density of bituminous specimens by gamma rays
EN 12697-8 Bituminous mixtures—Test methods for hot mix asphalt—Part 8: Determination of void characteristics of bituminous specimens
EN 12697-9 Bituminous mixtures—Test methods for hot mix asphalt—Part 9: Determination of the reference density

EN 12697-10 Bituminous mixtures—Test methods for hot mix asphalt—Part 10: Compactability

EN 12697-11 Bituminous mixtures—Test methods for hot mix asphalt—Part 11: Determination of the affinity between aggregate and bitumen

EN 12697-12 Bituminous mixtures—Test methods for hot mix asphalt—Part 12: Determination of the water sensitivity of bituminous specimens

EN 12697-13 Bituminous mixtures—Test methods for hot mix asphalt—Part 13: Temperature measurement

EN 12697-14 Bituminous mixtures—Test methods for hot mix asphalt—Part 14: Water content

EN 12697-15 Bituminous mixtures—Test methods for hot mix asphalt—Part 15: Determination of the segregation sensitivity

EN 12697-16 Bituminous mixtures—Test methods for hot mix asphalt—Part 16: Abrasion by studded tyres

EN 12697-17 Bituminous mixtures—Test methods for hot mix asphalt—Part 17: Particle loss of porous asphalt specimen

EN 12697-18 Bituminous mixtures—Test methods for hot mix asphalt—Part 18: Binder drainage

EN 12697-19 Bituminous mixtures—Test methods for hot mix asphalt—Part 19: Permeability of specimen

EN 12697-22 Bituminous mixtures—Test methods for hot mix asphalt—Part 22: Wheel tracking

EN 12697-23 Bituminous mixtures—Test methods for hot mix asphalt—Part 23: Determination of the indirect tensile strength of bituminous specimens

EN 12697-24 Bituminous mixtures—Test methods for hot mix asphalt—Part 24: Resistance to fatigue

EN 12697-25 Bituminous mixtures—Test methods for hot mix asphalt—Part 25: Cyclic compression test

EN 12697-26 Bituminous mixtures—Test methods for hot mix asphalt—Part 26: Stiffness

EN 12697-27 Bituminous mixtures—Test methods for hot mix asphalt—Part 27: Sampling

EN 12697-28 Bituminous mixtures—Test methods for hot mix asphalt—Part 28: Preparation of samples for determining binder content, water content, and grading

EN 12697-29 Bituminous mixtures—Test methods for hot mix asphalt—Part 29: Determination of the dimensions of a bituminous specimen

EN 12697-30 Bituminous mixtures—Test methods for hot mix asphalt—Part 30: Specimen preparation by impact compactor

EN 12697-31 Bituminous mixtures—Test methods for hot mix asphalt—Part 31: Specimen preparation by gyratory compactor

EN 12697-32 Bituminous mixtures—Test methods for hot mix asphalt—Part 32: Laboratory compaction of bituminous mixtures by vibratory compactor

EN 12697-33 Bituminous mixtures—Test methods for hot mix asphalt—Part 33: Specimen prepared by roller compactor

EN 12697-34 Bituminous mixtures—Test methods for hot mix asphalt—Part 34: Marshall test

EN 12697-35 Bituminous mixtures—Test methods for hot mix asphalt—Part 35: Laboratory mixing

EN 12697-36 Bituminous mixtures—Test methods for hot mix asphalt—Part 36: Determination of the thickness of a bituminous pavement

EN 12697-37 Bituminous mixtures—Test methods for hot mix asphalt—Part 37: Hot sand test for the adhesivity of binder on precoated chippings for HRA

EN 12697-38 Bituminous mixtures—Test methods for hot mix asphalt—Part 38: Common equipment and calibration

EN 12697-39 Bituminous mixtures—Test methods for hot mix asphalt—Part 39: Binder content by ignition

EN 12697-40 Bituminous mixtures—Test methods for hot mix asphalt—Part 40: In situ drainability

EN 12697-41 Bituminous mixtures—Test methods for hot mix asphalt—Part 41: Resistance to deicing fluids

EN 12697-42 Bituminous mixtures—Test methods for hot mix asphalt—Part 42: Amount of coarse foreign matter in reclaimed asphalt

EN 12697-43 Bituminous mixtures—Test methods for hot mix asphalt—Part 43: Resistance to fuel

EN 13043 Aggregates for bituminous mixtures and surface treatments for roads, airfields, and other trafficked areas

EN 13108-1 Bituminous Mixtures—Material specifications—Part 1: Asphalt concrete

EN 13108-2 Bituminous mixtures—Material specifications—Part 2: Asphalt concrete for very thin layers

EN 13108-3 Bituminous mixtures—Material specifications—Part 3: Soft asphalt

EN 13108-4 Bituminous mixtures—Material specifications—Part 4: Hot rolled asphalt

EN 13108-5 Bituminous mixtures—Material specifications—Part 5: Stone mastic asphalt

EN 13108-6 Bituminous mixtures—Material specifications—Part 6: Mastic asphalt

EN 13108-7 Bituminous mixtures—Material specifications—Part 7: Porous asphalt

EN 13108-8 Bituminous mixtures—Material specifications—Part 8: Reclaimed asphalt

EN 13108-20 Bituminous mixtures—Material specifications—Part 20: Type testing

EN 13108-21 Bituminous mixtures—Material specifications—Part 21: Factory production Control.

EN 13179-1 Tests for filler aggregate used in bituminous mixtures—Part 1: Delta ring and ball test

EN 1427 Bitumen and bituminous binders—Determination of the softening point—Ring and ball method

EN 196-6 Methods of testing cement—Determination of fineness

ÖNORM B 3584 2006 Asphaltmischgut—Mischgutanforderungen Splittmastixasphalt. Regeln zur Umsetzung der ÖNORM EN 13108-5, Ausgabe 1006-12-01

prEN 12697-44 Bituminous mixtures - Test methods for hot mix asphalt Part 44: Crack propagation by semi-circular bending test

prEN 12697-46 Bituminous mixtures - Test methods for hot mix asphalt Part 46: Low temperature cracking and properties by uniaxial tension tests

SIST 13108-5:2008 Bituminous Mixtures—Material Specifications—Part 5: Stone mastic asphalt—Requirements—Rules for implementation of SIST EN 13108-5

SN 670103a:2005 Gesteinskörnungen für Asphalt und Oberflächenbehandlungen für Straßen, Flugplätze und andere Verkehrsflächen

ZN-71/MK-CZDP-3 Drogi samochodowe. Uszorstnienie nawierzchni masą mineralno-asfaltową o wysokiej zawartości grysów (in Polish)

Index

A

Active grains, 10
Added fillers, 29
Adhesion promoters, 72
Aggregates
  abrasion value test (AAV), 52
  all-in, 53–54, 258
  artificial (slag), 71–72, 232
  coarse, 9, 11–12, 15, 57–60, 82–88, 92–94
  coating, 11, 49–50, 98, 123, 128, 130,
    136–137, 198, 224
  compacted skeleton part, 110
  crushing, 11, 49–50, 98, 123, 128, 130,
    136–137, 198, 224
  EN standards for, 258
  finalizing changes in, 88
  fine, 21–23
  gradation limits applying, designing
    advantages, 80
    disadvantages, 80
    gradation curve, 81
    larger than 2 mm, 81
    less than 2 mm, 98
  impact resistance, 23
  particle shape, 27, 130, 226
  properties
    coarse, 49
    fillers, 49
    fine, 49
  requirements for, 49
    according to European Standard
      EN 13043, 51–52
    system based on EN 13043, 53–56
    in United States, 65–66
  skeleton
    defined, 9
    passive and coarse grains,
      filling voids, 22
  surface area, 24, 100
  types of, 49
Aggregates abrasion value (AAV) test, 52
Airfield pavements
  Gardermoen airport in Oslo, 242
  Johannesburg airport
    porous mix, 242
    trial sections of, 242
  requirements for
    FAA and FOD, 239–240

  macrotexture, measurements of, 240
  PSV index, 240
  specification guidelines, 239–240
SMA
  airport in Frankfurt on Main,
    241–242
  Fisher–Tropsch wax, 241
  northern runway,
    replacement, 241–242
  pros and cons of, 240–241
  runway, 241
Air voids, 124
American Association of State Highway and
  Transportation Officials (AASHTO),
  18–19, 22–23, 46, 50, 70, 95,
  108–109, 111, 115, 122, 137, 140–141,
  144–145, 224, 232–233
Amplitude of vibration, 178, 181
Antiglare properties, 217
Antinoise properties
  asphalt mixture, impact of, 229
  noise, defined, 228
  test results
    close-proximity (CPX) method, 229
    noise levels of SMA, 230
    pavement noise increases, 228
    SMA, macrotexture of, 228–229
    sound-absorbing power, 228
Antiskid properties
  pavement qualities, 230
  PSV, 230
  SMA, macrotexture depth, 230–231
  SMA wearing courses
    grit, 232
    macrotexture, 232
    microtexture, 232
    mixtures with lower maximum
      aggregate size, 232
  test results
    AC macrotexture, 231
    Argentinean investigations, 231–232
    cement concrete macrotexture, 231
    gritting use, 231
    SCRIM, 232
    slurry seal macrotexture, 231
    SMA layer, PSV and final skid-resistance
      relationship between, 232
    SMA macrotexture, 231
    surface dressing macrotexture, 231

295